biogeography and
plate tectonics

FURTHER TITLES IN THIS SERIES

1. A.J. Boucot
   EVOLUTION AND EXTINCTION RATE CONTROLS

2. W.A. Berggren and J.A. van Couvering
   THE LATE NEOGENE – BIOSTRATIGRAPHY, GEOCHRONOLOGY AND
   PALEOCLIMATOLOGY OF THE LAST 15 MILLION YEARS IN MARINE AND
   CONTINENTAL SEQUENCES

3. L.J. Salop
   PRECAMBRIAN OF THE NORTHERN HEMISPHERE

4. J.L. Wray
   CALCAREOUS ALGAE

5. A. Hallam (Editor)
   PATTERNS OF EVOLUTION, AS ILLUSTRATED BY THE FOSSIL RECORD

6. F.M. Swain (Editor)
   STRATIGRAPHIC MICROPALEONTOLOGY OF ATLANTIC BASIN AND
   BORDERLANDS

7. W.C. Mahaney (Editor)
   QUATERNARY DATING METHODS

8. D. Janóssy
   PLEISTOCENE VERTEBRATE FAUNAS OF HUNGARY

9. Ch. Pomerol and I. Premoli-Silva (Editors)
   TERMINAL EOCENE EVENTS

Developments in Palaeontology and Stratigraphy, 10

biogeography and plate tectonics

J. C. Briggs

Department of Marine Science, University of South Florida, St. Petersburg, Florida, U.S.A.

ELSEVIER
Amsterdam — Oxford — New York — Tokyo   1987

ELSEVIER SCIENCE PUBLISHERS B.V.
Sara Burgerhartstraat 25
P.O. Box 211, 1000 AE Amsterdam, The Netherlands

Distributors for the United States and Canada:

ELSEVIER SCIENCE PUBLISHING COMPANY INC.
655, Avenue of the Americas
New York, NY 10010, U.S.A.

First edition 1987
Second impression 1988

ISBN 0-444-42743-0 (Vol. 10)
ISBN 0-444-41142-9 (Series)

Printed in The Netherlands

PREFACE

But if rivers come into being and perish and if the same parts of the earth are not always moist, the sea also must necessarily change correspondingly. And if in places the sea recedes while in others it encroaches, then evidently the same parts of the earth as a whole are not always sea, nor always mainland, but in process of time all change.

Aristotle, *Meteorologica*, ca. 335 B.C.

Our modern living world, the biosphere, may be subdivided into a number of biogeographic regions and provinces, each with its own distinctive complex of species. An important goal of research is to become better acquainted with the history of these various biogeographic units, for the composition of the ecosystem in each is a reflection of its past. We find, that as time has gone on, the relationship of the biota of the various units to one another has changed and that such changes may often be correlated with the gradual geographical alteration of the earth's surface. The historical approach to biogeography not only helps us to understand the biological effects of the geological changes but often sheds additional light on the geological events themselves. Perhaps most important, the more we learn about the interrelationship between historical biology and geology, the better we understand the evolutionary process.

Not long ago, Jardin and McKenzie (1972), in a brief overview of the biological effects of continental drift (plate tectonics), observed that the facts of continental drift had become so firmly established that it was no longer profitable for biologists to speculate about the past arrangements of land masses. In a similar vein, van Andel (1979) stated that the reconstruction of paleogeography can be carried on based only on physical data without recourse to paleobiogeographical evidence; he noted further that the physical world of the past, thus resurrected, can be used to interpret the biological one without the danger of circular reasoning. If these enthusiastic remarks were indeed true, the task of biogeographical research would be greatly simplified!

This attempt to provide information about continental relationships based on biological evidence to compare with geophysical data, is made with the realization that our lack of knowledge about the history of the various groups of animals and plants is difficult to overcome. At the family level, certainly fewer than one percent of the groups can be said to be reasonably well known in a systematic sense. In the final analysis, our knowledge about the evolution and geographical distribution of families and higher categories depends on competent systematic work. However, relatively little of this kind of research is being done. It is paradoxical, that, on one hand, we are so dependent on the systematist (including those who work with fossil as well as recent materials) for the facts about evolutionary relationship yet, on the other hand, systematics is considered by many to be old fashioned and unworthy of support. If we are to continue to improve our knowledge about the biological history of the earth, it is vital that systematic research be continued.

In analyzing distributional patterns and relating them to continental drift, it is important to attempt to separate effects of drift from various kinds of migration. As is noted in this book, most predrift relationships are very old in a biological sense. For example, Madagascar-India probably separated from Africa, and Euramerica was apparently cut off from Asia, in the mid-Jurassic. By late Jurassic/early Cretaceous times, South America departed from Africa and Africa from Euramerica. In evaluating the evolutionary effects of such events, it is necessary to consider phylogenetic relationships at the level of order, suborder, or family.

Although it is clear that the rate of speciation is quite variable, it is probably safe to say that most living species are not over five million years old and that the great majority of modern genera are Tertiary in origin, making them less than 65 million years old. Most of the families in such relatively well known groups as the birds, mammals, and flowering plants are not older than Cretaceous (65 – 130 million years) in age. This means that for widespread species and genera and for some families we should look for relatively recent (Tertiary) means of dispersal rather than attempting to invoke continental movement that took place in the Mesozoic. Claims that continental drift was responsible for the separation of extant species (Ferris et al., 1976; Platnick, 1976; Tuxen, 1978) are particularly suspect.

Since we know so little about the phylogeny of the various widespread groups of plants and animals, it is important to take advantage of all the information that does exist. The most complete analysis of terrestrial biogeography currently available was based on vertebrate animals only and was published 29 years ago (Darlington, 1957). When one adds the more recent information about the land and freshwater vertebrates, plus the results of systematic work on terrestrial and freshwater invertebrates and plants, and finally data on the distribution of some marine plants and animals, it is possible to obtain a better, if still woefully incomplete, idea of the history of oceanic and continental relationships.

One needs to look at only a small portion of the enormous literature on plate tectonics that has been published in the last 15 years to realize that there are many differences among the various reconstructions that have been presented. It becomes obvious that, although there is a general agreement about the presence of an assembly of continents (a Pangaea) in the early Mesozoic, there is considerable disagreement among earth scientists as to the configuration of the assembly and the manner and timing of the subsequent dispersal. While the revolution in geophysics was taking place, systematic work in paleontology and neontology was going on. There now is a need to incorporate this biological evidence into the theory of plate tectonics.

In order to understand the biological effects of the continental disbursement that took place beginning in the early Mesozoic, it is important to set the stage by first reviewing the consequences of continental assembly. Although the Permian/Triassic boundary has been recognized for many years as a time of severe extinction in the fossil record, the magnitude of this event was not fully appreciated until an analysis was made by Raup (1979). Using data on well-skeletonized marine vertebrate and invertebrate animals, he determined the percent extinction for the higher taxonomic groups. Then, using a rarefaction curve technique, he calculated the percent of species extinction that must have been responsible for the disappearance of the higher groups. His results indicated that as many as 96% of all marine species may have become extinct.

Although the fossil data pertaining to terrestrial forms are not plentiful enough to permit a direct comparison, there is little doubt that extensive extinctions took place there also. Padian and Clemens (1985) noted a sharp drop in the generic diversity of terrestrial vertebrates at the end of the Permian. The coming together of continental faunas that have developed in isolation for a long time may be expected to result in an extensive loss of species. The best documented example took place when North and South America were joined in the late Pliocene by the rise of the Isthmus of Panama (Simpson, 1980; Marshall, 1981; Webb, 1985b). The great losses caused by this event, especially in South America, prompted Gould (1980) to remark that it must rank as the most devastating biological tragedy of recent times.

Why did so many animals (and presumably plants) die out all of a sudden at the end of the Permian? In the marine environment, as the various continents closed with one another, the total amount of shore line and the associated continental shelf habitat (where the marine species diversity is the greatest) became greatly reduced. This restriction was undoubtedly accompanied by a loss of marine provinces (Schopf, 1980). A concurrent event was a significant drop in the salinity of the world ocean. Many salt deposits accumulated in isolated ocean basins that were being closed during the Permian (Flessa, 1980). Most marine species are quite stenohaline and would not be able to survive a significant drop in salinity. Stevens (1977) estimated that the accumulation of salt deposits during the Permian was equal to at least 10% of the volume of salt now in the oceans. But Benson (1984) maintained that this salinity reduction was not enough to cause a general reduction of the normal marine faunas.

In the terrestrial environment, in addition to the major loss almost certainly due to continental linkage, the advent of a severe continental climate associated with the assembled continents would cause further losses (Valentine and Moores, 1972). One may conclude that the coalition of continents, which resulted in the formation of the Triassic supercontinent of Pangaea, was a disastrous event for the world's biota. It was, in fact, the greatest catastrophe ever recorded. It took the world millions of years to recover the diversity that had existed in the early Permian. Additional, but less drastic, extinctions have taken place since the Permian/Triassic event. There is some evidence that these may have occurred at approximate 26 Ma intervals (Raup and Sepkoski, 1984) but there are no indications that these are attributable to plate tectonics.

In 1977, Smith and Briden devoted an entire volume to a series of Mesozoic and Cenozoic paleocontinental maps so that students, teachers, and research workers could use them to plot their own paleogeographic, paleontologic, or paleoclimatic data. The maps were computer drawn based on the input of geophysical data by the authors. These maps, while providing the outlines of the major continental blocks, gave no indication of the position of ancient shore lines and thus no separation between the terrestrial and marine environments.

An attempt to remedy the situation was made by Barron et al. (1981) by the production of a series of "paleogeographic" maps covering the same time period. They drew a distinction between paleocontinental maps, defined as those based on geophysical data, and paleogeographic maps which also utilized fossil and other sedimentary data. In their maps, ancient shore lines are depicted allowing the maps to be more useful for paleoclimatic and paleobiogeographic purposes. However, even though they represent a significant advance, the maps by Barron et al. (1981) need to be improved in order to accurately reflect the continental and oceanic relationships that are indicated by fossil and contemporary biological data.

Another atlas of continental movement maps, covering the past 200 million years, was published by Owen (1983). This work provided two series of maps, one assuming an earth of constant modern dimensions with the second assuming an earth expanding from a diameter of 80% of its modern mean value 180 – 200 million years ago to its modern size. While the expanding earth concept appears to solve some difficulties in the fit of the continental blocks, the technique is basically that of taking the continents in their modern dimensions and moving them about on the globe. There is no consideration of changes brought about by continental accretion or eustatic variation in sea level. Consequently, the use of these maps for biogeographical purposes is very limited.

The idea that we live on a world in which the geographical relationships of the continents are constantly changing has had a far reaching effect. It has not only caused a revolution in the earth sciences but it has stimulated the biological sciences and the public imagination. Hundreds of articles have appeared in the popular literature and even school children are sometimes introduced to continental drift as a part of their beginning geography. In both the scientific and popular press, the concept of Pangaea and the drift sequences tend to be depicted in a positive manner which does not indicate that our knowledge about such things is still very fragmentary.

It is particularly important to attempt to obtain dependable information about certain critical times in the history of continental relationships. We need to know when the terrestrial parts of the earth were broken apart and when they were joined together. The present investigation makes it clear that we cannot depend entirely on evidence from plate tectonics nor will purely biological evidence suffice. The world of the geophysicist is different from that of the biologist and unfortunately there is very little contact between the two camps.

This work represents an attempt to correlate biological events with the general history of continental movement. The biological data include information on many widespread groups of plants and animals. The intercontinental relationships of each group is of value to the overall scheme but the various groups are seldom easily comparable. Each group has its own age, evolutionary rate, area of origin, and dispersal ability. In some, such as certain mammalian orders and families, there is sufficient fossil evidence to help provide a fairly complete look into the past, but for the great majority, fossils are scarce or absent. For all the biotic groups, systematic works which attempted to reconstruct the evolutionary history were of great value. The result has been the accumulation of a large mass of data which by themselves are not very meaningful but when put together provide important insights into the course of continental relationships.

Since the general acceptance of the theory of plate tectonics, there have been published a number of papers on individual groups of organisms in which the authors have interpreted modern patterns in terms of the past relationships of the continents. However, there has been no comprehensive effort to relate to continental movement evidence about the biogeography of many, widespread groups of organisms. As such, this work represents a new departure in the study of biogeography. Also, almost all previous books on the subject have attempted to depict ancient distributional events on modern world maps. That practice needs to be abandoned. In this work, if there are indications that the major part of a distributional pattern was established at a given time in the past, it is depicted on a map appropriate to that time.

A continuing difficulty in the pictorial presentation of continental drift is that most published illustrations have been made using some kind of lateral projection that give an equatorial view of the earth. The distortions inherent in such projections become greatly magnified when one is attempting to illustrate events that took place in the high latitudes of the globe. It is more useful and realistic to use projections that utilize the equal area concept and also show both poles. The accompanying series of maps (see Appendix) use the Lambert equal-area type of projection and attempt to provide outlines of land and sea that appear to be indicated by our present knowledge of biology and geophysics.

ACKNOWLEDGEMENTS

The bibliographic research that eventually led to this book got underway in 1980/81 wen I was on sabbatical leave at Stanford University. At that time, the work was supported by a grant from the National Aeronautic and Space Administration (no. NAG 2 – 74). The project was carried on and the manuscript completed during 1981/1986 at the University of South Florida. I wish to thank Daniel F. Belknap, Richard Estes, and Pamela Hallock Muller for their helpful comments. I am indebted to Carole L. Cunningham and Jodi S. Gray for their expert secretarial help.

CONTENTS

INTRODUCTION: THE DEVELOPMENT OF THE SCIENCE

> The first appearance of animals now existing can in many cases be traced, their numbers gradually increasing in the more recent formations, while other species continually die out and disappear, so that the present condition of the organic world is clearly derived by a natural process of gradual extinction and creation of species from that of the latest geological periods.
>
> Alfred R. Wallace, *On the Law Which has Regulated the Introduction of New Species,* 1855

For the past 20 years, the time during which the geophysical concept of continental drift has become fully accepted, there has developed a need for biogeographers to take a fresh look at their discipline in the light of past changes in the relationships of the land masses and oceanic basins of the world. As the new plate tectonic framework becomes adopted, biogeography will undergo a change from an emphasis on modern distributional patterns to a greater appreciation for the historical development of such patterns.

In order to realize the importance of the new plate tectonic approach, one should take the time to place it in the context of significant changes that have occurred in the past. As is true of many disciplines, unless one is familiar with its historical progression, one cannot appreciate its present position in the stream of events, nor predict its future course.

IN THE BEGINNING

In the 17th century, the task of biogeographers was a relatively simple one. The book of Genesis told how all men were descended from Noah and that they had made their way from Armenia to their present countries. Since there had been a single geographical and temporal origin for man, the consensus was that this was also true for all animals and that they had a common origin from which they too had dispersed (Browne, 1983). So scholars like Athanasius Kircher (1602 – 1680) and his contemporaries set themselves the task of working out the details of the structure of the Ark so that it could accommodate a pair of each species of animal. It is interesting to see that this exercise of deducing the structure, and eventual grounding place, of the Ark has been repeated dozens of times in the past 300 years. In the year of 1985, there were news reports of five different expeditions busily combing the slopes of Mt. Ararat for the remains of the Ark.

Since well before Kircher's time, travelers and explorers had been bringing back to Europe thousands of specimens representing unknown species of animals. As these were described, secular scholars were obliged to find room for them aboard the Ark. No one seemed to have worried about the thousands of species of plants that could not have survived the Deluge. By the time the 18th century arrived, the idea of the Ark had to be abandoned by people who were informed on the subject of natural history. However, the concept of the Deluge was still strongly entrenched so that a reasonable substitute for the Ark had to be found.

The person who came to the rescue was a young man in Sweden named Carl Linnaeus (1707 – 1778). He was a deeply religious person who felt that God spoke most clearly to man through the natural world. In fact, it has been said that Linnaeus considered the universe a gigantic museum collection given to him by God to describe and catalogue into a methodical framework (Browne, 1983). Linnaeus proceeded to solve the Ark problem by telescoping the story of the Creation into that of the Deluge. He proposed that all living things had their origin on a high mountain at about the time the primeval waters were beginning to recede. Furthermore, he proposed that this Paradisical mountain contained a variety of ecological conditions arranged in climatic zones so that each pair of animals was created in a particular habitat along with other species suited for that place.

As the flood waters receded, Linnaeus envisioned the various animals and plants migrating to their eventual homes where they remained for the rest of time. For him, species were fixed entities that stayed just as they were created. In other works, Linnaeus emphasized that each species had been given the structure that was the most appropriate for the habitat in which it lived. This insistence on a close connection between each species and its habitat, exposed Linnaeus to criticism by other scholars. How could the reindeer, which was designed for the cold, have made its way across inhospitable deserts to get from Mt. Ararat to Lapland?

The Comte de Buffon (1707 – 1788), who published his great encyclopedia, *Histoire Naturelle* in 1749 – 1804, was influential in persuading educated people to give up the Garden of Eden concept and also the idea that species did not change through time. He apparently believed that life originated generally in the far north during a warmer period and had gradually moved south as the climate got colder. Because the New and Old Worlds were almost joined in the north, the species in each area were the same. But, as the southward progression took place, the original populations were separated. In the New World, some kind of a structural degeneration took place which caused those species to depart from the primary type. In regard to mammals, Buffon observed that those of the New and Old World tropics were exclusively confined to their own areas. This has been subsequently referred to as "Buffons Law" and interpreted to mean that such animals had evolved in situ and had not migrated from Armenia (Nelson, 1978).

As the result of the influence of Buffon and others, the idea of a single biblical center for all species was replaced by the idea of many centers of creation, each species in the area where it now lived (Browne, 1983). This, and the Linnaean concept of the importance of species as identifiable populations that existed in concert with other species, encouraged naturalists to think in terms of groups of species characteristic of a given geographic area. Linnaeus and his students and others began to emphasize the contrasts among different parts of the world by publishing various "floras" and "faunas". Johannes F. Gronovius published his *Flora Virginica* in 1743; Carl Linnaeus his *Flora Suecica* in 1745, *Fauna Suecica* in 1746, and *Flora Zeylandica* in 1747; Johann G. Gmelin his *Flora Sibirica* in 1747 – 1769; and Otto Fabricius his *Fauna Groenlandica* in 1780.

From the viewpoint of the mid-18th century, it may be seen that biogeography underwent a fundamental change during the preceding 100 years. Naturalists were at first occupied with the problems of accommodation aboard the Ark and the means by which animals were able to disperse the various parts of the world following the Deluge. The Ark concept gave way to the Paradisical mountain which in turn yielded to the idea of creation in many different places. At the same time, the Linnaean axiom of the fixity of species through time was replaced by one of change under environmental influence. Finally, naturalists began to study the associations of plants and animals in various parts of the world and, in so doing, began to appreciate the contrasts among different countries.

Johann Reinhold Forster (1729 – 1798) was a German naturalist who emigrated to England in 1766. From 1770 to 1772 he published several small works including a volume entitled *A Catalogue of the Animals of North America.* In 1772, he together with his son Georg, was given the opportunity to accompany Captain Cook on his second expedition to the South Seas. This was a three-year circum-navigation of the globe. Upon his return, Forster published his *Observations made during a Voyage round the World* in 1778. In this work, he presented a worldwide view of the various natural regions and their biota. He described how the different floras replaced one another as the physical characteristics of the environment changed. He also called attention to the way in which the type of vegetation determined the kinds of animals found in each region.

Forster compared islands to the mainland and noted that the number of species in a given area was proportionate to the available physical resources. He remarked on the uniform decrease in floral diversity from the equator to the poles and attributed this phenomenon to the latitudinal change in the surface heat of the earth. He found the tropics to be beautiful, rich, and enchanting – the area in which nature reached its highest and most diversified expression (Browne, 1983). Forster, more than any of his predecessors, understood that biotas were living communities characteristic of certain geographical areas. Thus the concept of natural biotic regions was born.

As knowledge of the organic world increased and greater numbers of species became known, naturalists tended to specialize in the study of either plants or animals. For some reason, it was the early botanists who took the greatest interest in biogeography. Karl Willdenow (1765 – 1812) was a plant systematist and head of the Berlin Botanical Garden. In his 1792 book *Grundriss der Krauterkunde*, he outlined the elements of plant geography. He recognized five principal floras in Europe and, like Forster, was interested in the effect of temperature on floral diversity. To account for the presence of the various botanical provinces, Willdenow envisioned an early stage of many mountains surrounded by a global sea. Different plants were created on the various peaks and then spread downward, as the water receded, to form our present botanical provinces.

Willdenow's most famous student was Alexander von Humboldt (1769 – 1859). Von Humboldt has often been called the father of phytogeography (Brown and Gibson, 1983). In his youth he was impressed and influenced by his friendship with Georg Forster. Von Humboldt felt that the study of geographical distribution was scientific inquiry of the highest order and that it could lead to the disclosure of fundamental natural laws (Browne, 1983). He became one of the famous explorer-naturalists and devoted much of his attention to the tropics of the New World. As a part of his great 24 volume work *Voyage aux Régions Equinoxiales du Nouveau*

Continent (1805 – 1837, with A.J.A. Bonpland), von Humboldt included his *Essai sur la Géographie des Plantes* (1805). The latter work, his best contribution to biogeography, was inspired as the result of his climbing Mt. Chimborazo, an 18,000-foot peak in the Andes. There he observed a series of altitudinal floral belts equivalent to the tropical, temperate, boreal, and arctic regions of the world.

The next significant step in the progress of biogeography was made by a Swiss botanist named Augustin de Candolle (1778 – 1841). In 1820, he published his important *Essai élémentaire de Géographie botanique.* In that work he made a distinction between "stations" (habitats) and "habitations" (the major botanical provinces). De Candolle was also the first to write about the notion of competition or a struggle for existence, noting that individuals competed for space, light, and other resources. De Candolle's work had a significant influence on such important figures as Charles Darwin, Joseph Hooker, and his own son Alphonse. The elder de Candolle was a close friend of von Humboldt and was surely influenced by him.

THE GEOLOGICAL CONNECTION

The study of extinct floras got underway with the work of Adolphe Brongniart who published his *Histoire des Végétaux fossiles* in 1828. He was followed by Alphonse de Candolle. Both men believed that life first appeared as a single primitive population evenly distributed over the entire surface of the globe. This uniform population was supposed to have gradually fragmented into many diverse groups of species (Browne, 1983). In the meantime, Georges Cuvier had begun his work on fossil vertebrates and many others soon followed. From a distributional standpoint, the first effective connection between fossil and contemporary patterns was made by Charles Lyell (1797 – 1875). In his *Principles of Geology* (1830 – 1832 and subsequent editions), Lyell undertook extensive discussions on botanical geography, including the provinces of marine algae, and on the geographical distribution of animals. In addition, he analyzed the effects of climatic and geological changes on the distribution of species and the evidence for the extinction and creation of species.

As Browne (1983) has pointed out, Lyell's suggestion that the elevation and submersion of large land masses resulted in the conversion of equable climats into extreme ones, and vice versa, according to the quantity of land left above sea level, was most important. This view meant that floras and faunas had to be dynamic entities capable of expanding or contracting their boundaries as geological agents altered topography and climates. So Lyell, the champion of gradual change to the earth's surface, brought to biogeography a sense of history and the realization that floral and faunal provinces had almost certainly been altered through time.

Edward Forbes (1815 – 1854), despite his short life, made important contributions to both terrestrial and marine biogeography. He accounted for the evident relationship between the floras of the European mountain tops and Scandanavia by supposing very cold conditions and land subsidence in the recent past. His map of the distribution of marine life together with a descriptive text that appeared in Alexander K. Johnston's *The Physical Atlas of Natural Phenomena* (1856) was the first comprehensive work on marine biogeography. In it, the world was divided into 25 provinces located within a series of 9 horizontal "homoizoic belts". A series of five depth zones was also recognized. In the same year, Samuel P. Woodward, the famous malacologist, published part three of his *Manual of the Mollusca* which dealt with the worldwide distribution of that group.

In 1859, Forbes posthumous work *The Natural History of European Seas* was published by Robert Godwin-Austen. In this work Forbes observed that (1) each zoogeographic province is an area where there was a special manifestation of creative power and that the animals originally formed there were apt to become mixed with emigrants from other provinces, (2) each species was created only once and that individuals tended to migrate outward from their center of origin, and (3) provinces to be understood must be traced back like species to their origin in past time. Another important contribution was made by James D. Dana who participated in the United States Exploring Expedition, 1838 – 1842. Through observations made on the distribution of corals and crustaceans, he was able to divide the surface waters of the world into several different zones based on temperature and used isocrymes (lines of mean minimum temperature) to separate them. His plan was published as a brief paper in the *American Journal of Science* in 1853.

The first attempt to include all animal life, marine and terrestrial, in a single zoogeographic scheme was by Ludwig K. Schmarda in his volume entitled *Die Geographische Verbreitung der Tiere* (1853). He divided the world into 21 land and 10 marine realms. However, it was P.L. Sclater who divided the terrestrial world into the biogeographic regions that, essentially, are still in use today. This was done in 1858 in a small paper entitled *On the General Geographical Distribution of the Members of the Class Aves*. Despite the fact that his scheme was based only on the distributional patterns of birds, Sclater's work proved to be useful for almost all groups of terrestrial animals. This has served to emphasize that biogeographic boundaries, found to be important for one group, are also apt to be significant for many others.

EVOLUTIONARY BIOGEOGRAPHY

When the young Charles Darwin visited the Galapagos Islands in 1835, he was struck by the distinctiveness, yet basic similarity, of the fauna to that of mainland South America. When Alfred Russel Wallace traveled through the Indo-Australian Archipelago, some 20 years later, he was puzzled by the contrasting character of the island faunas, some with Australian relationships and others with southeast Asian affinities. After considerable thought about such matters (many years on Darwin's part), each man arrived at a theoretical mechanism (natural selection) to account for evolutionary change. The key for both Darwin and Wallace was the realization that distributional patterns had evolutionary significance.

The announcement of their joint theory by Darwin and Wallace in 1858 in the *Journal of the Linnean Society of London* and, especially, the publication of Darwin's *Origin of Species* in 1859, changed the thinking of the civilized world. Darwin included two important chapters on geographical distribution in his book. In discussing biogeography from the viewpoint of evolutionary change, Darwin made three important points: (1) he emphasized that barriers to migration allowed time for the slow process of modification through natural selection; (2) he considered the concept of single centers of creation to be critical; that is, each species was first produced in one area only and from that center it would proceed to migrate as far as its ability would permit; and (3) he noted that dispersal was a phenomenon of overall importance.

In regard to the third point, Darwin observed that oceanic islands were generally volcanic in origin and must have accumulated their biota by dispersal from some mainland source. He felt that the presence of alpine species on the summits of widely separated mountains could be explained by dispersal having taken place during the glacial period when such forms would have been widespread. More important, he suggested that the relationships that biologists were then finding between the temperate biotas of the northern and southern hemispheres were attributable to migrations made through the tropics during the glacial period when world temperatures were cooler. Finally, he noted that the preponderant interhemispheric migratory movement had been from north to south and suggested that this was due to the northern forms having advanced through natural selection and competition to a higher stage of dominating power.

When Darwin was going through the long process of formulating his theory, his closest confidants were Charles Lyell and Joseph D. Hooker. Hooker, a great plant collector and systematist, having accompanied Sir James Ross on his Antarctic Expedition (1839 – 1843), was particularly interested in southern hemisphere botany. Hooker felt that Darwin was perhaps too dependent on dispersal in accounting for disjunct relationships. In describing the flora of New Zealand in 1853, Hooker speculated on the possibility that the plants of the Southern Ocean were the remains of a flora that had once been spread over a larger and more continuous tract of land than now exists in that part of the world. In modern terms, he was suggesting a vicariant rather than a dispersal history for the subantarctic floras.

While Darwin went on to investigate many other aspects of evolutionary change, Wallace applied himself primarily to biogeography. Finally, in 1876, Wallace published his monumental two volume work *The Geographical Distribution of Animals*. In that work, he reached a number of conclusions about biogeography that are still worth reviewing. For example, he pointed out that (1) paleoclimatic studies are very important for analyzing extant distribution patterns; (2) competition, predation, and other biotic factors play important roles in the distribution, dispersal, and extinction of animals and plants; (3) discontinuous ranges may come about by extinction in intermediate areas or patchiness of habitats; (4) disjunctions of genera show greater antiquity than those of a single species, and so forth for higher categories; (5) the common presence of organisms not adapted for long distance dispersal is good evidence of past land connections; (6) when two large land masses long separated are reunited, extinction may occur because many organisms will encounter new competitors; (7) islands may be classified into three major categories, continental islands recently set off from the mainland, continental islands long separated from the mainland, and oceanic islands of volcanic and coralline origin; and (8) studies of island biotas are important because the relationships among distribution, speciation, and adaptation are easier to see and comprehend.

Wallace did considerable traveling in the Indo-Australian region and was particularly concerned about the location of the dividing line between the Oriental and Australian faunas. As George (1981) has noted, Wallace, by 1863, had decided that the line should run from east of the Philippines south between Borneo and Celebes and then between Bali and Lompok. It was illustrated in his 1876 work and later in his book *Island Life* in 1880. Although Wallace, in his 1910 book *The World of Life,* changed his mind about the affiliation of Celebes, his original line is the one generally called "Wallace's Line". It is represented on his regional scheme (Fig. 1) which is close to that proposed earlier by Sclater.

Following the publication of Wallace's works, many biogeographers repeated his distribution plan without any major new interpretations. In 1890, E.L. Trouessart published his *La Géographie zoologique* which examined both terrestrial and marine patterns. In 1895, Frank E. Beddard came out with *A Text-book of Zoogeography*. In 1907, Angelo Heilprin published a volume entitled *The Geographical and Geological Distribution of Animals.* The latter introduced some minor changes to Wallace's map and also reviewed the information then available about the distribution of fossil forms. Also a number of works, dealing with the establishment of hypothetical land bridges and the rise and fall of mid-ocean continents, were published. But, as our knowledge of sea-floor history increased, these theories were discarded.

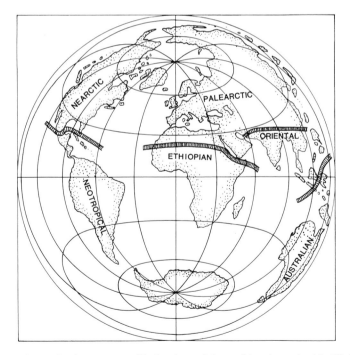

Fig. 1. The six zoogeographical regions of the world as determined by Wallace in 1876. This scheme has stood the test of time and has proved useful for many widespread groups of animals and plants.

The next significant advance in biogeography took place in 1915 when William Diller Matthew (1871 – 1930), a geologist and paleontologist, published his article on *Climate and Evolution*. Matthew was an expert on fossil mammals and his 1915 work was devoted primarily to emphasizing the importance of the northern hemisphere (the Holarctic Region) in the evolution and dispersal of that group. However, the most important aspect of that work has turned out to be Matthew's statement of his theory about centers of dispersal. He said, "At any given period, the most advanced and progressive species of the race will be those inhabiting that region; the most primitive and unprogressive species will be those remote from this center. The remoteness is, of course, not a matter of geographic distance but of inaccessibility to invasion, conditioned by the habitat and facilities for migration and dispersal."

Progress in our knowledge about distribution patterns in the marine environment was made by Arnold Ortmann when he published his *Grundzüge der Marinen Tiergeographie* (1896). The following year, in 1897, Philip L. Sclater published a paper on the distribution of marine mammals. In 1935, Sven Ekman completed the huge task of analyzing all of the pertinent literature on marine animal distribution and published his results in a book entitled *Tiergeographie des Meeres*. In 1953, a second edition was printed in English. Modern books on marine zoogeography have been published by John C. Briggs, *Marine Zoogeography* (1974), Geerat J. Vermeij, *Biogeography and Adaptation* (1978), S. van der Spoel and A.C. Pierrot-Bults (eds.), *Zoogeography and Diversity in Plankton* (1979), and Oleg G. Kussakin (ed.) *Marine Biogeography* (1982, in Russian).

In the 1920s and 1930s a new development took place which combined the rapidly evolving field of ecology with biogeography. The beginning was marked by the appearance of Friedrich Dahl's *Grundlagen einer ökologischen Tiergeographie* in 1921 and Richard Hesse's *Tiergeographie auf ökologischer Grundlage* in 1924. These efforts were apparently in response to a need to examine the geographical distribution of plant and animal communities on a local and worldwide scale. A revised English edition of Hesse's book was prepared by W.C. Allee and Karl P. Schmidt and published in 1937. This was followed by a second edition in 1951. Other works that have carried on this approach are Marion I. Newbigin's *Plant and Animal Geography* published in 1936, V.G. Gepner's *General Zoogeography* (1936, in Russian), Frederic E. Clements and Victor E. Shelford's *Bio-ecology* in 1939 (which introduced the biome concept), and the work by L.R. Dice *The Biotic Provinces of North America* in 1943. Among such works, that of Robert H. MacArthur and Edward O. Wilson, *Island Biogeography* (1967), deserves special mention. Its explanation of the relationship between colonization and extinction and its analysis of the species-area concept, had a stimulating impact on both biogeography and ecology. Other modern examples of the combined approach are the books by P.M. Dansereau, *Biogeography; An Ecological Perspective* (1957), Brian Seddon, *Introduction to Biogeography* (1971); C. Barry Cox, Ian N. Healey, and Peter D. Moore, *Biogeography* (1973); and James H. Brown and Arthur C. Gibson, *Biogeography* (1983).

In 1944, a significant work on phytogeography, *Foundations of Plant Geography,* was published by Stanley A. Cain. His analyses of fossil distributions and his discussion of the center of origin concept have been most useful to later workers. A work of similar importance for those interested in the distribution of animals was published by Philip J. Darlington, Jr. in 1957. Although *Zoogeography: the Geographical Distribution of Animals* was based only on patterns demonstrated by the terrestrial and freshwater vertebrates, it represented an important milestone because it was the first time in the 20th century that all of the information about those animal groups had been gathered together. Since the data on fossil vertebrates are, in general, better than those for the invertebrate groups, Darlington's book had great significance for historical biogeography.

Darlington (1957) emphasized that the major worldwide patterns of vertebrate animals indicated a series of geographical radiations from the Old World tropics. Such radiations were considered to take place because competitively dominant animals were continually moving out from their tropical centers of origin. In a later article, Darlington (1959) observed, "The history of dispersal of animals seems to be primarily the history of successions of dominant groups, which in turn evolve, spread over the world, compete with and destroy and replace older groups, and then differentiate in different places until overrun and replaced by succeeding groups."

THE ADVENT OF CONTINENTAL DRIFT

It was not until the late 1950s that the idea of historic continental movement began to be taken seriously by large numbers of earth scientists. Much earlier, between 1910 and 1912, Frederick B. Taylor, H.D. Baker, and Alfred L. Wegener had all advanced views about continental drift similar to those that are held today. However, at that time, the earth's crust was almost universally considered to have a solid structure without movement.

Between 1915 and 1929, Wegener published four editions of his book *Die Entstehung der Kontinente und Ozeane* including an English edition (*The Origin of Continents and Oceans*). These works created considerable controversy but most geologists and geophysicists were still not convinced. Research into paleomagnetism then began to offer some supporting evidence for drift. In 1960, Harry H. Hess made the suggestion that the sea floors crack open along the crest of the mid-ocean ridges, and that new sea floor forms there and spreads apart on either side of the crust. Robert S. Dietz named this process sea-floor spreading and coupled with it the suggestion that old sea floor is absorbed beneath zones of deep ocean trenches and young mountains.

J. Tuzo Wilson (1963, 1973) noted that oceanic islands tended to increase in age away from the mid-ocean ridges and that certain "hot spots" existed where strings of volcanic islands had been formed. These and other discoveries led to the modern view of plate tectonics which holds that the earth's crust is divided into a mosaic of shifting plates in which the continents are embedded. We now have available many reconstructions of continental relationships covering the last 200 million years.

The plate tectonic revolution in earth science had a gradual but decisive effect on biogeography. Previously, it had been necessary to discuss the historical relationships of the biogeographical regions and their biotas within the framework of stable continents. Now that biologists were released from this constraint, there were varied reactions. In 1965, Darlington published his book *Biogeography of the Southern End of the World*. He was able to contrast the life of southern South America, southern Africa, India, Australia, New Zealand, and Antarctica, and decided that these lands had once been situated much closer together. Darlington also noted that successive new groups of plants and animals had been invading the southern ends of the world over a long period of time. But counterinvasions from south to north were exceedingly rare.

In 1969, Miklos D.F. Udvardy published his *Dynamic Zoogeography* which emphasized the importance of dispersal under different climatic conditions but did not attempt to assess continental drift. Several important paleontological works, such as *Faunal Provinces in Space and Time* (F.A. Middlemiss and P.F. Rawson, eds., 1971), *Organisms and Continents Through Time* (N.F. Hughes, ed., 1973), and *Atlas of Paleobiogeography* (A. Hallam, ed., 1973), took drift into consideration. E.C. Pielou's textbook *Biogeography*, published in 1979, devoted a chapter to continental drift. P. Banarescu in his *Principii si Probleme di Zoogeografie* (1970) did the same. Also a number of brief overviews on the biological effects of drift were published as journal articles; for example, Jardine and McKenzie (1972), Raven and Axelrod (1972, 1974), and Cracraft (1975).

THE RISE OF VICARIANISM

The most important and controversial development of the decade of the 1970s was the enthusiastic promotion of the theory of "vicarianism". Vicariance refers to the biogeographic patterns produced by a particular kind of allopatric speciation in which a geographic barrier develops so that it separates a formerly continuous population. This distinguishes vicarianism from the kind of allopatric speciation which takes place as the result of migration or dispersal of individuals across an existing barrier to colonize the other side. Although these two kinds of allopatric speciation had been recognized for many years, the advocates of vicarianism came to feel that their viewpoint had been neglected and the vicarianism was *the* important process in producing evolutionary change.

Vicarianism got its start at the American Museum of Natural History in New York. Its original promoters had read and become impressed by the publications of Leon Croizat, a man who had produced voluminous works written in a wandering and confused style that almost defied analysis. But Croizat was hailed as a newly discovered genius and became the Patron Saint of vicarianism (Nelson and Rosen, 1981). A connection between vicarianism and plate tectonics was established by envisioning, before the separation of the continents, a "hologenesis", a kind of primitive cosmopolitanism based on a theory espoused by Rosa (1923). Hologenesis, where species were supposed to have been created with cosmopolitan ranges, may be contrasted with the center of origin concept (Darwin, 1859) where species originated in a limited area and then spread as far as their capabilities would permit.

As their enthusiam for a supposedly new concept (which actually may be traced back to the works of Adolph Brongniart and Alphonse de Candolle) grew, the proponents of vicarianism emphasized that it was really vicariance that produced geographical differentiation and multiplication of species while dispersal produced only sympatry. In the best explanation of the mechanics of vicarianism, Croizat et al. (1974) stated, "The existence of races or subspecies that are separated by barriers (vicariance) means that a population has subdivided, or is subdividing, not that dispersal has occurred, or is occurring across the barriers." Belief in vicariance led its disciples to maintain that centers of origin do not exist since, to recognize such centers, they would have to concede that species are capable of dispersing from their places of origin to establish themselves elsewhere, the usual result being, after a period of time, allopatric speciation by migration rather than by geologic change. Consequently, Croizat et al. said, "We reject the Darwinian concept of the center of origin and its corollary, dispersal of species, as a conceptual model of general applicability in historical biogeography."

In the late 1970s and early 1980s many journal articles were published about the pros and cons of vicarianism. In 1981, two books appeared, one edited by Gareth Nelson and Donn E. Rosen, *Vicariance Biogeography: A Critique,* and the other written by Gareth Nelson and Norman Platnick, *Systematics and Biogeography.* It has been implied that one must use the vicarianist approach if one is to examine distributions in the light of continental drift and that the biogeographical regions of Wallace and Sclater are no longer useful (Nelson and Platnick, 1980). We have been told that the endemism apparent at various oceanic islands of the Pacific can be explained by vicarianism rather than by dispersal (Springer, 1982).

In the meantime, before vicarianism had gotten underway, a book by Willi Hennig, *Phylogenetic Systematics* (1966), was published. This was the second edition of a book originally published in German in 1950. By the 1970s, this work began to have a significant impact on the methodology employed by people who did systematic work. Hennig provided a set of rules for the practice of systematics which have collectively been called "cladism". These rules have generally been helpful but the one that applies to biogeography has turned out to be suspect. It states that species possessing the most primitive characters are found within the earliest occupied part of the area, i.e., the center of origin for that group. Although this rule was at first enthusiastically adopted by some, very little biogeographical evidence has been found to support it.

From the vantage point of the middle 1980s, it may be said that there are some interesting signs of shifts in position. McCoy and Heck (1983) said vicarianists now admit that allopatric speciation via dispersal can take place and only maintain that it is less important than vicariance. Cracraft (1983) had indicated that some cladists can forego their center of origin concept in order to join forces with the vicarianists. Judging from the number of recent articles that have employed cladograms (phylogenetic diagrams) along with diagrams illustrating geographic relationships, this seems to be true.

The modern case for the center of origin concept has been stated by Briggs (1984a) in a monograph entitled *Centres of Origin in Biogeography.* The main conclusions reached in that work are:

(1) Information now available suggests that centers of origin are evident in the

oceans, the freshwaters, and the terrestrial environments of the earth. For the more advanced orders and families, the centers are located in the tropics. The characteristics of such centers are large geographic size, heterogeneous topography, warm and relatively steady temperatures, maximum species diversity for the general part of the world in which they are located, and possession of the most advanced species and genera of those groups of organisms that are well represented.

(2) On a worldwide basis, the study of major barriers that separate one biogeographic region from another tells us that species produced in the centers can not only spread out to occupy large portions of the regions in which they evolved, but can sometimes transgress the barriers and colonize adjacent regions. As this process goes on, a given center may eventually have a profound influence on the composition of the flora and fauna of a large portion of the world. Evolutionary centers, because of their high levels of species diversity and possession of the more advanced and more highly competitive species, have a very high resistance to invasion by species from other areas.

(3) The kind of evolution that goes on in the centers is probably different than that which takes place in areas peripheral to such centers. Evidently large populations in which the individuals possess high levels of genetic variation are involved. Parapatric speciation and the kinds of allopatric speciation that permit natural selection to operate in large populations are probably important. The rate of evolutionary change is bound to be slower than that which occurs in small, isolated populations. But, in terms of producing continuing phyletic lines, it is probably more successful.

(4) The data pertaining to centers of origin and their probable mode of operation indicate that we live on a world in which some parts, in terms of evolutionary progress, have been considerably more important than others. The complicated community structure and species relationships of the highly diverse tropical areas are not well understood, yet much of the biota of these areas is in the process of being destroyed for agricultural and other purposes. The primary goal of international conservation should be the preservation of significant portions of the tropical ecosystems, both terrestrial and marine. In an evolutionary sense, these areas represent the future of the living world.

A recent work, which continues the crusade of denigrating the importance of dispersal while extolling the virtues of vicarianism, is that by Humphries and Parenti (1986). These authors consider dispersal biogeography to be an unscientific, ad hoc discipline that " . . . can never let us discover the history of the earth." In contrast, vicariance hypotheses are regarded as scientific because they are testable. It is stated that two tests may be applied to a vicariance hypothesis: add more tracks (reinforcement by other taxa that show the same pattern) and compare the hypothesis with a geological one. It must here be emphasized that any biogeographic hypothesis based only on the distribution and relationships of a single group of organisms is on shaky ground. The strongest hypotheses are those based on common patterns demonstrated by many different biotic groups and are, at the same time, consistent with a well substantiated geological history. It makes no difference whether the hypothesis involves vicariance or dispersal or both. Such "tests" (if they really can be considered as such) are certainly not the exclusive property of the vicariance method.

THE PRESENT WORK

This book has been written from the point of view that dispersal and vicarianism have each played an important role in historical biogeography. Most higher organisms have, as an integral part of their life history, a dispersal phase which allows them to spread out and occupy new territories. This enables each species to eventually move as far as its migratory ability and ecological versatility will permit. As a young species enlarges its territory, it will encounter barriers that will prohibit or delay its further expansion. Such barriers may comprise one or more of a variety of physical, chemical, and biological features of the environment.

It is important to realize that barriers to migration are not often static and that, over time, most of them have been or will be changed. Sometimes the creation of a barrier will result in the interruption of the range of a species or a species complex. For example, when the Isthmus of Panama was finally completed at about the end of the Pliocene, it separated the tropical marine biota of the New World into two parts, one inhabiting the Eastern Pacific and the other the Western Atlantic. This is considered to be a vicariant event, in that it prevented gene flow between the two parts and caused the separated populations to embark on their own evolutionary courses. But, at the same time, the isthmian connection provided a dispersal corridor between North and South America for terrestrial and freshwater organisms, also with profound evolutionary (and ecological) consequences.

In a similar manner, when the land connection across the Bering Strait was first made in the late Cretaceous, it separated the marine populations of the Bering Sea – Arctic Ocean but connected terrestrial North America to Asia. From a marine standpoint, the creation of Beringia was a vicariant event but from a terrestrial viewpoint it provided a dispersal opportunity. The tectonic uplift of a mountain range can constitute an important barrier for lowland forms but simultaneously may present a migratory corridor for species of the high-altitude biota.

For the past 15 years, a significant portion of the theoretical literature on biogeography has been devoted to argument about the efficacy of vicarianism compared to dispersalism. It is important that biogeographers attempt to appreciate the biosphere as a whole instead of concentrating too heavily on a single habitat.

Dispersal is an everyday occurrence undertaken by succeeding generations of almost all species while vicarianism is an event of much greater rarity since it must involve the creation of a barrier to separate existing populations. A most important point is that when vicariance does take place it appears to offer, at the same time, unusual dispersal opportunities for some groups of species. So dispersion may be looked upon as a continuing, inexorable process while vicariance, when it occurs in one habitat usually stimulates dispersal in another. This is particularly true in regard to continental movement with its making and breaking of land and sea barriers.

Part 1

THE NORTHERN CONTINENTS

Chapter 1

THE NORTH ATLANTIC CONNECTION

> It is this union of passionate interest in the detailed facts with equal devotion to abstract generalisation which forms the novelty in our present society. This balance of mind has now become part of the tradition which infects cultivated thought. It is the salt that keeps life sweet.
>
> Alfred North Whitehead, *Science and the Modern World,* 1925

A detailed reconstruction of the paleobathymetry of the Atlantic Ocean from the Jurassic to the present was published by Sclater et al. (1977). They paid particular attention to the area north of 35°N where they felt that previous studies had been inadequate. They proposed a sequence of events beginning with a tight fit of the North Atlantic continents in the Jurassic about 125 million years ago. At that time, the main oceanic body of the North Atlantic was evident but apparently no motion had occurred among the land masses comprising northern Europe, Greenland and North America. At 95 Ma, it was assumed that Europe had started to separate from North America and that Greenland had moved eastward far enough to open up the Labrador Sea.

A more recent article on the same subject by Sclater and Tapscott (1979) was illustrated in greater detail and depicted the presence of continuous sea passages between Europe and North America, including both sides of Greenland, from 165 Ma to the present. The maps of Barron et al. (1981) show Europe and North America continuously separated beginning 160 Ma. However, in his review of plate movements and their relationship to biogeographic changes, Hallam (1981) referred to a persistent land connection between Greenland and Europe that lasted until the end of the Eocene (38 Ma). What did actually happen? Can the biological data help to resolve this question?

In regard to the marine fauna of the North Atlantic, its obvious impoverishment (Briggs, 1970) compared to that of the North Pacific, has puzzled biologists for a long time. An early explanation was that a land bridge across the North Atlantic must have blocked off its connection to the Arctic and North Pacific Oceans. This theory was apparently first published by Forbes (1859) who, in turn, gave credit to Sir John Richardson for suggesting this explanation. Forbes described the bridge as probably extending from 70° to 75°N and completing in its northern coastline the symmetrical form of the Arctic Basin. Prior to the concept of plate tectonics, other North Atlantic land bridges of various configurations were proposed by several authors.

Biological evidence, brought forth within the past 12 years, is of considerable help. Work on the fossil mammals (terrestrial) of the early Tertiary has been particularly useful. For the late Paleocene, Kurtén (1973) found a strong relationship between the faunas of Europe and North America and considered them to belong to a single zoogeographic region. But, by the late Eocene, this resemblance had largely disappeared and the two areas were judged to lie in different zoogeographic regions. Other mammalian evidence (McKenna, 1975; West and Dawson, 1978) is consistent with Kurtén's conclusions and appears to indicate the presence of a migration corridor at the beginning of the Eocene which was probably interrupted by a water gap sometime during the early Eocene. Thiede (1980) concluded that the presence of late Paleocene to early Eocene marine fossils on Svalbard (Spitsbergen) indicated the first opening of a shallow-water seaway between the Arctic Ocean and the North Atlantic.

Most recently, Savage and Russell (1983) observed that the early Eocene in Europe witnessed a mass immigration of land mammals from North America. As a result, about 50 – 60% of the genera of mammals from the Sparnacian (Lower Eocene) Stage of the Paris Basin are the same as genera that have been found in the Wasatch strata of Wyoming. At about the time the immigrants arrived, some 80% of the genera known from the late Paleocene of Europe became extinct. Representatives of several lizard families (helodermatids, varanids, gekkonids, and agamids) apparently also took the same route at about the same time (Estes, 1983).

The salamanders (urodele amphibians) demonstrate some interesting transatlantic relationships. The North American genera of the family Salamandridae are considered to be a derived subgroup of a predominantly Eurasian family. It has been suggested that they originated in Europe and dispersed to North America in the early Cenozoic (Milner, 1983). The family Proteidae has living genera in southeast Europe and eastern North America. Fossil proteids are known from the Upper Paleocene of North America and from the Miocene of southwest Russia and Germany. An extinct related family, the Batrachosauroididae, is known from the Cretaceous to the Miocene of North America and the Paleocene and Eocene of Europe. Both families, therefore, inhabited Euramerica prior to the early Eocene division of that continent. Unlike the salamandrids, they may have arisen in North America.

The distribution patterns of some of the freshwater fishes are useful. Within the *oculatus* species group of fossil and recent gars (Lepisosteidae), there is a European – North American relationship that Wiley (1976) attributed to an early Eocene continuity. Two species of bowfins (Amiidae) were common to North America and Europe in the early Tertiary (Boreske, 1974) and Eocene remains of gars and bowfins have recently been found on Ellesmere Island, northwest of Greenland (Patterson, 1981a). A study of the systematics of the family Percidae by Collette and Banarescu (1977) indicated that the family probably originated in Europe and then dispersed over a North Atlantic land route sometime between the end of the Cretaceous and the beginning of the Eocene. An Eocene interruption of contact between the two continents then allowed their percid faunas to develop independently. In North America, the percid tribe Ethiostomatini had undergone a remarkable evolutionary radiation resulting in three genera and about 150 species.

The North Atlantic connection must have been important for much of the plantlife of the Mesozoic and early Tertiary. A number of conifers and early angiosperms, that now exist only as relicts, once had broad holarctic distributions. Examples are such genera as *Ginko, Sequoia, Liriodendron,* and *Ceridiphyllum* (Axelrod, 1983). Fossils of these genera have been found on Greenland as well as either Spitsbergen or Iceland so the evidence for their once continuous distribution across the North Atlantic seems very good. Another example is the section *Aigeros* of the genus *Populus*. The species of this section do not occur in Asia but do have a disjunct distribution across the Atlantic.

In regard to the marine fauna of the North Atlantic, Fallow (1979) published a paper showing a strong, positive correlation between the width of the ocean basin and the degree of similarity of the invertebrate animals that inhabit the continental shelf on each side. His data, on the width of the ocean basin from the early Jurassic to Neogene times, were taken from the plate tectonic work of Sclater et al. (1977). In this case, the earth sciences data has produced a broad outline of the evolution of the North Atlantic Basin that is consistent with the information from marine biology. However, it is evident that general studies of plate movements have not been able to focus with sufficient accuracy on the details of the relationship among the terrestrial areas of Europe, Greenland and North America. Thus, the studies of Sclater and Tapscott (1979) and Barron et al. (1981) showed these areas to be separated by sea passages from the mid-Jurassic to the present while the biological data, based on the distributions of terrestrial mammals, amphibians, reptiles, shallow marine fossils, and freshwater fishes, indicate that land connections must have persisted from the Mesozoic into the Cenozoic as late as the early Eocene.

Additional information that has a bearing on this problem became available in 1983 with the publication of the proceedings of a NATO Advanced Research Institute held in Italy in 1981. The meeting was devoted to the structure and development of the Greenland-Scotland Ridge. There were two papers dealing with paleontology (Hoch, 1983; McKenna, 1983a). Both recognized the presence of two early Tertiary land bridges. One, called the Thulian route, extended from Labrador and Baffin Island through Greenland, The Faeroe Islands and Scotland. The other, the DeGeer route, connected Ellesmere Island, Greenland, Spitsbergen, and Scandanavia. A geophysical account (Nunns, 1983) appeared to show that the separation of the Greenland area from Europe began in the early Paleocene.

Chapter 2

THE NORTH PACIFIC CONNECTION

> Widely ranging species, abounding in individuals, which have already triumphed over many competitors in their own widely extended homes will have the best chance of seizing on new places, when they spread into new countries.

Charles Darwin, *The Origin of Species*, 1859

As the tectonic plates on each side of the North Atlantic began to separate, those of the North Pacific area began to draw closer to one another. By the late Cretaceous (80 Ma), the maps of Smith and Briden (1977) still show a large gap between Asia and North America but those of Barron et al. (1981) indicate that the two continents have made contact. Also, Fujita (1978) described a series of Cretaceous collisions among several small plates between Siberia and North America.

Lillegraven et al. (1979) found fossil evidence of land vertebrate exchanges in the region of Beringia which were determined to have followed a continental collision that took place in the late Cretaceous. The Rocky Mountain, Bug Creek Fauna of that time has revealed new mammalian genera (multituberculate and placental) which have been recognized as probable immigrants, presumably from Asia (Webb, 1985a). This Bering land connection, once established, evidently endured for a long time instead of being interrupted in the early Cenozoic as indicated by Barron et al. (1981). Durham and McNeil (1967) were unable to find any evidence for the migration of marine invertebrates between the North Pacific and the Arctic – North Atlantic in the early Tertiary. Also, during this time, the marine mammal faunas on each side of the land bridge were strikingly different; the desmostylians, sea lions, and ancestral walruses were confined to the North Pacific, while the true seals of the family Phocidae were found in the Arctic – Atlantic (Hopkins, 1967). Early Tertiary marine fossils from northern Alaska have been interpreted to indicate an isolation or near isolation of the Arctic Ocean that lasted from the end of the Cretaceous until sometime in the Eocene (Marincovich et al., 1985).

ANIMAL MIGRATIONS

The broad expanse of Beringia that emerged in the late Cretaceous provided an almost continuous highway for the dispersal of terrestrial and freshwater fauna that lasted through almost all the Tertiary and was again available during the glacial stages of the Pleistocene. Relatively brief inundations apparently took place in the late Pliocene, and during the interglacial stages (Hopkins, 1967; Herman and Hopkins, 1980). The biogeographic importance of the Beringia connection has been noted in works dealing with the distribution of mammals, reptiles, various insect groups, and freshwater fishes.

The extensive review work on the mammalian paleofaunas of the world by Savage and Russell (1983) and a chapter on mammalian diversification in North America by Webb (1985a) are most useful in determining the role of Beringia in the history of the Tertiary mammalian fauna of the northern hemisphere. In early Paleocene time there was a sudden (explosive) origin of orders, families, and genera. The order Condylarthra has been traced from one genus in the late Cretaceous Bug Creek fauna to 20 genera in the early Paleocene (Van Valen, 1978). The explosive early Paleocene interval may have generated the following five orders: Condylarthra, Cimolesta, Insectivora, Dermoptera, and Carnivora, in addition to the Arcto-cyonia, Primates, Leptictida, and Taeniodonta that had already appeared in the latest Cretaceous (Webb, 1985a). By the mid-Paleocene, three genera of mesonychids had appeared that represent the order Acreodi, which had been known from earlier deposits in Asia. As many as three genera of large herbivores, representing the order Pantodonta, also evidently immigrated into North America from an earlier origin in Asia (Simons, 1960). An important event of the late Paleocene was the appearance of three families in North America that showed relationships to both South America and Asia. All three groups, the edentates, notoungulates, and xenungulates, were best developed in South America. It has been suggested (Gingerich, 1985) that they originated in South America then migrated, in the late Paleocene, to North America and then to Asia via Beringia.

In the early Eocene, several genera, apparently immigrants from Asia, appeared in North America for the first time (Webb, 1985a). These include genera belonging to the orders Tillodontia, Rodentia, and Pantodonta. A little later in the early Eocene, the first perissodactyls, artiodactyls, adapid and omomyid primates, hyaenodontids, and tapiroids, all possible immigrants from Asia, appeared. In the mid-Eocene, the North American mammalian fauna became considerably more endemic. But, in the late Eocene, the Asiatic influence was renewed. The new immigrants appear to be adapted to woodland savanna and scrubby habitats (Webb, 1985a) and included several groups of lophodont rodents, pig-like entelodonts, several families of artiodactyls including the Camelidae, and several groups of perissodactyls including tapiroids, rhinocerotids, and chalicotheres.

With the onset of the Oligocene, came the most impressive faunal turnover in the whole age of mammals (Webb, 1985a). In the larger mammal groups, there was the first American appearance of the Canidae, Felidae, Mustelidae, Tapiridae, Castoridae, Rhinocerotidae, Anthracotheriidae, and Tayassuidae. Many of the new groups arrived by way of the Bering Strait. According to Webb, the groups most traceable to Asiatic stock are the Castoridae, Anthracotheriidae, and Tapiridae. A great wave of immigration from Asia came in the early Miocene when some 16 genera established themselves in North America. Among the more conspicuous forms were species of cats, bears, pronghorn antelopes, beavers, and flying squirrels. By the mid-Miocene, the proboscidian genera *Miomastodon* and *Gomphotherium* arrived along with crecitid rodents but, in general, this was a time when the Asian influence had become slowed. Finally, toward the end of the Miocene, there was a renewed surge of immigrants including several large carnivores, large ungulates, and small herbivores.

In the Pliocene, there occurred another burst of intercontinental migration resulting in an extensive faunal turnover at the generic level. Some 72 new genera appeared (Savage and Russell, 1983) most of which were probably of Asiatic origin.

The microtine rodents which include the meadow mice, muskrats, lemmings, and related forms, have a center of evolution and diversity in Asia (Repenning, 1980). Noteworthy invasions of North America appear to have taken place at about 5.3, 3.7, 1.8, 1.2, and 0.47 Ma. The cessation of intercontinental migrations between the invasion periods appears to be attributable to climatic and vegetation changes except for the period from 3.7 to 2.5 Ma. The latter is correlated with the Tertiary opening of the Bering Strait.

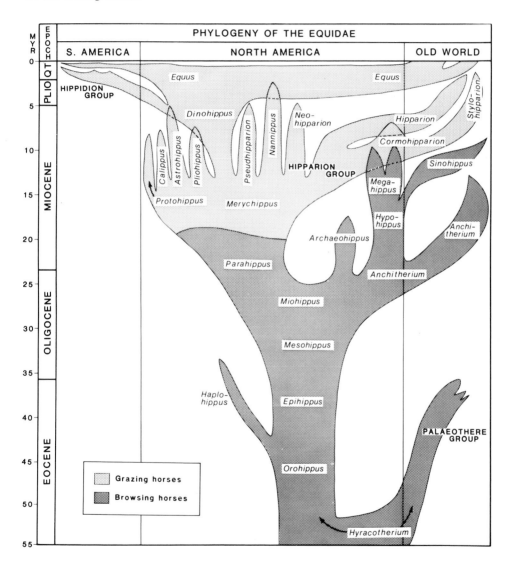

Fig. 2. A phylogeny of the Equidae indicating dispersal of several primitive genera to the Old World via Beringia. After MacFadden (1985).

24

The first Pliocene episode of large mammal intermigration took place about 3.5 – 3.7 Ma (Kurtén and Anderson, 1980). Migrants from the Old World to the New included such forms as a weasel, a raccoon, a bear, a cat, and a hyena. Early camels, horses, and cheetahs went the opposite direction. In regard to horses, the major evolutionary steps in this group clearly took place in North America. As these events occurred, a series of westward migrations across Beringia took place. These began in the Eocene and involved several primitive genera (MacFadden, 1985) (Fig. 2). In the Pleistocene, eight different species of the modern genus *Equus* made the crossing and several of them continued on to Africa (Bennett, 1980) (Fig. 3).

A second active phase of intermigration began at about 1.8 Ma when there was a influx of Eurasian forms such as jaguars, bovids, a mammoth, a caribou, and lemmings (Kurtén and Anderson, 1980). The third phase took place in the late Pleistocene when a host of forms from the Old World reached Alaska. Included were the muskox, moose, bison, a weasel, and a fox. *Homo sapiens* probably arrived during the most recent (Wisconsin) ice age. Although earlier mammal dispersals across Beringia took place in both directions, those of the late Pleistocene appear to have been entirely from west to east.

In regard to reptiles, it has been suggested that a primitive lizard group called the varanoid necrosaurs, together with the true varanids (Varanidae) may have originated in Asia and dispersed across the Bering connection in the late Cretaceous (Estes, 1983). Later, in the early Cenozoic, it is likely that three genera (*Eumeces, Scincella, Neoseps*) of the lizard family Scincidae reached North America from Asia. The family Crotalidae, the pit vipers, probably originated in southeast Asia

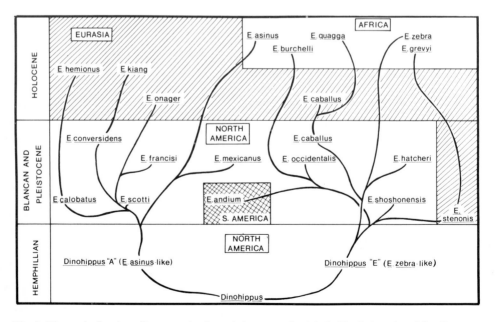

Fig. 3. Dispersal of various *Equus* species from their center of origin in North America. After Bennett (1980).

(Rabb and Marx, 1973). Two genera dispersed across the Bering Land Bridge, probably in the Miocene, and evolved into the modern rattlesnakes, water moccasin, copperhead, fer-de-lance, and bushmaster (Brattstrom, 1964).

The salamander family Plethodontidae, a large group of 23 genera and more than 200 species, has a predominantly North American distribution. However, two southern European species belonging to the genus *Hydromantes* are members of that family. Three other species of that genus are found only in California. This biogeographic puzzle has attracted the interest of several herpetologists. Wake et al. (1978) studied genetic variation and albumin evolution within *Hydromantes* and determined that the phylogenetic separation between the European and American species probably took place in the Oligocene. The migration must have taken place via the Bering Land Bridge since the North Atlantic connection had been broken much earlier.

Earlier salamander relationships involving Beringia have been proposed by Milner (1983). Living species belonging to the family Cryptobranchidae are found in East Asia and in the Appalachians of North America. Fossils are known from the Middle Oligocene to the Pliocene of Europe and from the Upper Paleocene onwards in North America. The family most closely related to the Cryptobranchidae is the Hynobiidae, a group with the majority of its species restricted to East Asia. Milner suggested that both families originated in Asia and that a cryptobranchid moved eastward across Beringia in the late Cretaceous or early Paleocene.

Insect migrations back and forth across Beringia must certainly have been numerous. Some interesting examples are: Among the caddisflies (Trichoptera), a primitive generic line (*Sortosa*) originated in Asia, migrated to North America in the late Cretaceous, then to South America where it produced a descendent line. The derived genus (*Chimarra*) then migrated to North America, went back across the land bridge to Asia in the Paleocene/Eocene, then in the late Eocene returned again to North America (Ross, 1958). The phylogenetic history of the fruit fly genus *Drosophila* reveals a series of radiations from the Old World tropics northward to the temperate zone then across to the New World. These migrations apparently involved more than 20 eastward crossings of Beringia (Throckmorton, 1975). Erwin (1981) described a carabid beetle lineage (*Trachypachus*) that originated in the New World and made an Eocene crossing to Asia.

The archaic freshwater fish family Polyodontidae (paddlefishes) is of considerable biogeographic interest for there are but two living species (each in a distinct genus), one in the Mississippi River system and one in the Yangtse system in China. Two extinct genera are known, one from the Eocene of Wyoming and one from the late Cretaceous of Montana (Patterson, 1981a). So, an Asian–North American relationship is indicated.

Several teleostean (bony fish) freshwater families that are represented in the Early and Middle Eocene Green River Formation of Wyoming, Colorado, and Utah, demonstrate transpacific relationships (Grande, 1985). The family Hiodontidae has only two living species in eastern North America but two Eocene species have been described from the Green River Formation. The closest relatives of the hiodontids are the Lycopteridae, an extinct family from eastern Asia. The fossil family Ellimmichthyidae is represented in the Green River Formation by *Diplomystus dentatus*. A closely related species has been described from China.

The family Clupeidae is represented in the Green River Formation by two valid species in the genus *Knightia* (Grande, 1985). A Chinese fossil belongs to the same genus. The family Catostomidae is represented in the Green River by one species whose closest relatives are found in Eocene to Recent deposits in Asia. Two species of the family Osteoglossidae are found in the Green River. Their closest relatives occur in the Eocene deposits of Australia and possibly Indonesia. Although modern osteoglossids are confined to freshwater, a Paleocene/Eocene fossil was a widespread marine species (Patterson, 1975), so the distribution of this family may not be a positive indication of migration via the Bering Land Bridge.

The pikes (Esocidae) are another teleost, freshwater group with a northern hemisphere history. The earliest fossils are from the Paleocene of North America and the Oligocene of Europe (Patterson, 1981a). There are five living species, one with a Holarctic distribution, one endemic to the Amur River basin in eastern Asia, and three confined to eastern North America. The presence of the oldest fossil and four of the five living species in North America indicates a probable origin in that area with subsequent dispersals to Asia. The species *Esox reicherti*, confined to the Amur River, may represent an early migration across Beringia and *E. lucius,* with its Holarctic range, a later one.

The mudminnows (Umbridae) are considered to be related to the pikes and are usually placed in the same suborder (Esocoidei). Their modern geographic pattern is similar to the pikes in that there are five living species with four of them occurring in North America. But the oldest fossil umbrid (*Palaeoesox*) has been found in the Eocene of Germany and another genus (*Proumbra*) occurs in the Oligocene of Siberia. The genus *Novumbra* occurs in the Oligocene of North America and is represented by a living species in the State of Washington. The genus *Dallia* occurs as a single living species in both Alaska and Siberia and *Umbra* has two living species in eastern North America and one in Europe. Thus the historical distribution of this family is not at all clear. It is probable that both the North Pacific and North Atlantic connections were involved.

Among the more advanced freshwater fishes, the catfishes of the family Ictaluridae probably represent an early invasion of North America from Asia. Assuming a southeast Asian origin for the Order Siluriformes (Briggs, 1979), the status of the Ictaluridae, an endemic family in North America, probably indicates that it has been on the latter continent for a long time. The one catfish family (Siluridae) that extends to northern Europe does not appear to be very close to the Ictaluridae. But, the Asian – African family Bagridae does appear to be closely related to the Ictaluridae and the present range of the bagrids extends north of the Amur River in Siberia (Berra, 1981).

Under slightly warmer climatic conditions, a crossing to North America via Beringia by a bagrid catfish, or by a form ancestral to both the Bagridae and the Ictaluridae, would have been possible. More recently, migrations by other families of freshwater fishes took place. The suckers (Catostomidae) and minnows (Cyprinidae) probably arose in Asia and reached North America via Beringia (Gilbert, 1976; Briggs, 1979). The genus *Catostomus* (Catostomidae), which originated in North America, has managed to invade the Asian side probably during the Pleistocene. Thus, we have evidence for two crossings of the land bridge by one freshwater family.

It was noted earlier (p. 18) that the freshwater fish family Percidae probably originated in Europe and first reached North America over a North Atlantic land route. During the Oligocene, when contact was established between Europe and Asia, two percid genera (*Perca* and *Sitzostedion*) apparently dispersed from Europe into Siberia. Later, in the Neogene, they reached North America by means of the Bering connection (Collette and Banarescu, 1977). So, it seems that various groups of freshwater fishes have moved back and forth across the Bering Land Bridge probably since the late Cretaceous.

PLANT MIGRATIONS

The subject of the relationship of the flora of temperate eastern North America to that of Japan and China has fascinated botanists of several generations. While early works by Linnaeus and some of his students remarked on a general relationship between elements of the Japanese and North American floras (Boufford and Spongberg, 1983), it was Asa Gray who first recognized a special affinity between the eastern North American and eastern Asian plants. Although Gray referred to this affinity in a series of papers from 1840 to 1878, his classic contribution was his analysis of the collection of Japanese plants made by Charles Wright during the U.S. North Pacific Exploring Expedition (Gray, 1859). It is interesting to note that Gray was encouraged to undertake his comparative study by Charles Darwin (Boufford and Spongberg, 1983). Gray not only called attention to the close systematic relationship of the two widely separated floras but suggested that an interchange between the two had taken place via Asia. Thus, he was the first to put forward the concept of a disrupted north temperate flora that was once more widely distributed.

Modern summaries of the eastern Asian – eastern North American relationship have been published (Li, 1952; Wood, 1972). In 1982, a symposium on the subject was held at the Missouri Botanical Garden and the papers given were published the next year. In discussing the two regions in general, White (1983) emphasized that most of the disjunct genera have more species in eastern Asia, that most of the disjunct families have more genera in eastern Asia, and that in eastern Asia there are more than four times as many species in disjunct families than there are in eastern North America. Cheng (1983) undertook a comparative study of two areas within the disjunct regions that have equivalent climates, Hubei Province in China and the Carolinas in the United States. He found that 75% of the families were shared as well as a large number of genera. In Hubei, the woody angiosperms were notably more diverse than in the Carolinas. The former possesses 73 families, 200 genera, and 650 species and the latter 46 families, 82 genera, and 217 species. It is within the woody angiosperm group that most of the disjunct genera are found.

Analyses of individual families and genera have provided important information about relative diversity and past migratory movements. In North America there are 23 genera and 106 species of orchids while eastern Asia has 80 genera and 350 species; more than two thirds of the genera exhibit phylogenetic ties between the two regions (Sing-chi, 1983). The maple genus *Acer* has about 200 species worldwide but three fourths of them occur in China (Ying, 1983). Wolfe (1981), through his ex-

28

amination of fossil material, has traced the development of *Acer* for the past 60 Ma. His analysis shows that the genus has a complicated history in which various species migrated back and forth across the Bering Land Bridge from the beginning of the Tertiary into the Quaternary (Fig. 4).

Similar histories to that of *Acer* are suggested by works on other plant groups. The genus *Magnolia* has eight species in the eastern U.S., 18 others southward into tropical America, and about 50 in eastern Asia; seven of the 11 sections within the genus are entirely Asian and in two of the American sections there are Asian species (Little, 1983). In the hawthorn genus *Crataegus,* Phipps (1983) suggested that the ancestral crossing of Beringia took place in the Eocene from China to the New World. This was apparently followed by additional eastward migrations in the Miocene and two possible subsequent migrations in the opposite direction. An in-

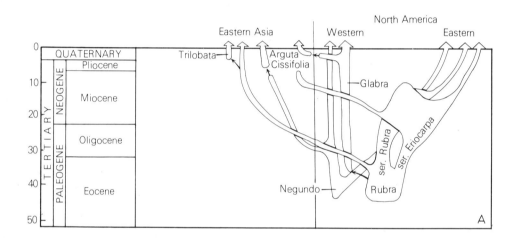

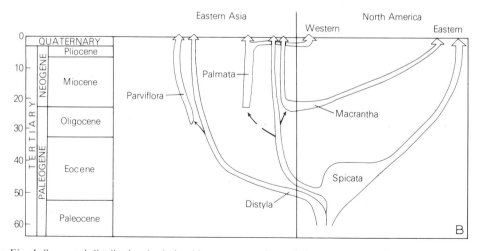

Fig. 4. Suggested distributional relationships among sections of the genus *Acer.* After Wolfe (1981).

vestigation of the family Scrophulariaceae by De-Yuan (1983) indicated that six genera probably migrated from eastern Asia to North America and that four genera must have taken the reverse route.

In regard to the chronology of plant dispersals involving the North Pacific connection, it seems that terrestrial migrations could have taken place beginning with the formation of the land bridge in the late Cretaceous. However, a dispersal across the entire Holarctic would not have been possible since Euramerica was separated from Asiamerica by two epicontinental seas, the Mid-Continental Seaway on one side and the Turgai Sea on the other (Map 5*). The presence of these barriers has been substantiated by palynological evidence of two floral provinces at the mid and high latitudes of the northern hemisphere (Muller, 1970; Hickey, 1984; Tschudy, 1984).

By the Paleocene, the Mid-Continental Seaway had become dried up allowing the formation of one supercontinent in the northern hemisphere (Map 6). This would have permitted, for the first time, a Holarctic distribution of the higher plants. For the angiosperms, it has been suggested that they originated in tropical southeast Asia in the early Cretaceous and then gradually spread to higher latitudes (Briggs, 1984a). From the Paleocene to the early Eocene, the primitive angiosperms and other plants could have moved out of Asia eastward across Beringia to North America and thence to Europe. But, in the early Eocene Europe became isolated by the opening of the North Atlantic Ocean (Map 7) so that Holarctic dispersal was interrupted.

It has been noted that mixed deciduous hardwood and conifer-hardwood forests occupied middle and high latitudes into the Middle Miocene (Axelrod, 1983). This complex has been called the Arcto-Tertiary flora. Following the mid-Miocene, the climate in the far north became colder and dryer and the forests were displaced to the south. In western North America, beginning near the end of the Miocene and continuing to the present time, almost the entire Rocky Mountain and adjacent areas were warped up as a large unit. Considerable erosion began to take place as well as faulting and volcanism. Along the Pacific coast, periods of marine deposition combined with uplifts resulted in a continental accretion that developed rapidly in the late Tertiary (Dott and Batten, 1971). This geological activity and associated climatic disturbances fragmented the flora of the west and resulted in high rates of species extinction. In the east, the Appalachian chain remained relatively quiescent and therefore retained much of its original species diversity.

In addition to western North America, the descendents of the Arcto-Tertiary flora in Europe are also a depauperate complex. This condition may be attributed to the isolation of the European continent during the Eocene (Map 7) and the existence of Tertiary epicontinental seas which flooded the lowlands. In the temperate part of the northern hemisphere, we have as our legacy today the rich forests and associated flora of eastern China and Japan, the related but less diverse complex in the mountains of eastern North America, and the species-poor communities of western North America and Europe.

---

* The maps are shown in the appendix at the end of the text.

THE PLEISTOCENE

Cenozoic glaciations have a much more extensive history than was at one time believed. We used to think of the Ice Age as being confined to the Pleistocene and consisting of four glacial periods. Instead, we now have evidence of northern hemisphere glaciation that goes back about 3 Ma and the Antarctic has been in a full ice age for about 15 Ma (van Andel, 1985). It is now possible to distinguish some 30 cold excursions separated by brief warm spells. Although plate movement has not been extensive during the Ice Age, it is probable that the Ice Age is primarily attributable to earlier changes in oceanic currents that were caused by plate tectonics.

At the beginning of the Tertiary, sea surface temperatures became gradually cooler at high latitudes. In the northern hemisphere the Arctic Ocean had been cut off from the North Pacific by late Cretaceous plate movement joining North America to Asia (Map 5). By the Paleocene (Map 6), the North American Mid-Continental Seaway dried up leaving only one shallow opening (the Turgai Sea) southward from the Arctic Ocean. In the early Miocene, the circum-global equatorial circulation was interrupted by the connection between Africa and Eurasia (Map 8). By the late Miocene, the Gibraltar portal was closed causing the desiccation of the Mediterranean and a lowering of the salinity of the oceans by about 6% (van Andel, 1985). This raised the freezing point of seawater so that sea ice could form more readily.

Although the passage between Greenland and Europe opened in the early Eocene (Map 7), and a shallow connection to the North Pacific has occurred intermittently since the Pliocene, the Turgai Sea dried up in the Oligocene and the Arctic Ocean has remained a relatively isolated deep basin. Another isolation effect took place in the southern hemisphere. This was caused by the opening, in the late Oligocene, of Drake Passage to the extent that a strong circum-Antarctic current could develop. This insolated the Antarctic continent from the warmer seas to the north. As the high-latitude climates continued to deteriorate, the onset of the Ice Age was probably aided by such isolation effects. Once established, the Ice Age has gone through many glacial-interglacial fluctuations. These phenomena are probably controlled by changes in the orbital behavior of the earth according to the Milankovitch hypothesis (Imbrie and Imbrie, 1979).

Although attention has been called to the Pleistocene interchange of mammalian faunas that occurred across Beringia (p. 24), many other elements of the boreal fauna and flora must have been involved. Each time the ice sheets built up, the sea level dropped and the shallow waters of the Bering Strait receded. The exposure of the land bridge presented, for the terrestrial biota, opportunities for intercontinental migration. At the same time, the land barrier separated the marine biota of the Arctic Ocean from that of the Bering Sea. These sea-level changes had important biogeographical effects in other parts of the world as well. As the level dropped, large islands in southeast Asia became connected to the mainland, New Guinea was attached to Australia, Tasmania to Australia, and the Red Sea was cut off from the Indian Ocean.

The ice sheets themselves excluded almost all animal and plant life from the areas they occupied. Species of the Arctic biota had to retreat to the south or occupy certain ice-free refugia (or become extinct). Associated climatic changes took place in many parts of the world with sometimes important biogeographic consequences. Compared to earlier Periods in the earth's history, the Quaternary has occupied the attention of many biologists, geologists, and paleontologists. The large volume of modern literature available makes it unnecessary to examine this Period in detail. Besides, to do so would require a large volume in itself.

Chapter 3

THE CARIBBEAN CONNECTION

Paleogeography starts as a concoction of essential ingredients, generally too meager, and winds up as a heady essence distilled through the imagination of the perpetrator.

W.P. Woodring, *Caribbean Land and Sea Through the Ages,* 1954

ANTILLES

Biologists and geologists have attempted to account for the presence of the Antillean terrestrial and freshwater biota by proposing three different methods of access: (1) a land bridge (or bridges) from the mainland to the islands; (2) movement of the islands themselves, with at least a partially intact biota, eastward from a position adjacent to the mainland; and (3) fortuitous overseas transport over a long period of time.

The land bridge hypothesis apparently began with Spencer (1895) who envisioned a gigantic Antillean continent that was connected to the mainland in the Pliocene and Pleistocene. This concept was strongly reinforced by Barbour (1916) who felt that it was impossible for many of the Antillean animals to have arrived by over-water dispersal. Other advocates were Scharff (1922), Schuchert (1935), and de Beaufort (1951). Rivas (1958) was convinced that the Cuban poeciliid fishes (a secondary freshwater group) had reached that island by means of a land connection to the Yucatan Peninsula.

The island-rafting hypothesis began with a paper by Rosen (1976) who presented a vicariance model of Caribbean biogeography. In this work, which was based primarily on the relationships of the freshwater fish faunas, Rosen proposed that many of the present Antillean species, or their immediate ancestors, were once part of a late Mesozoic fauna that occupied a "Proto-Antilles" archipelago. The action of the East Pacific plate that supposedly pushed the Proto-Antilles eastward to become the modern Antilles was considered to be a vicariant event, in that the islands so transported were supposed to have carried their biota with them.

In providing geophysical data to back up his theory, Rosen (1976) gave Malfait and Dinkelman (1972) credit for being the main architects of a plate tectonic model of Caribbean evolution. According to Rosen, this model provided that North and South America were connected, in the late Jurassic, by an archipelago (in the present location of Central America) that was a forerunner to the present Antillean archipelago. This was an error for the paper by Malfait and Dinkelman did not refer to nor illustrate the presence of such a Proto-Antilles chain.

In a more recent paper on Caribbean biogeography, Rosen (1985) attempted to provide, " . . . a precise means of specifying how a given biohistory is explicitly tied to a particular geohistory." In order to accomplish this goal, he advised using branching diagrams (cladograms) that supposedly demonstrate congruence between cladistic messages from biology with cladistic messages from geology. Non-cladistic solutions to the problem of Caribbean history were dismissed as being imprecise. Although cladograms are often a useful way to express relationships, equivocal data are not improved by referring to them as cladistic messages and presenting them in the form of cladograms.

In the period 1980 – 1982, several geophysical studies were published suggesting that a proto-Greater Antilles may have existed in the late Cretaceous/early Tertiary as an arc of volcanic islands between Mexico and South America (Hedges, 1983). These islands were then supposedly swept eastwards by plate movement to become the modern Greater Antilles.

This surge of interest in the tectonic history of the Caribbean region has continued to the present time. Geophysical works by Pindell and Dewey (1982), Sykes et al. (1982), Burke et al. (1984), Duncan and Hargraves (1984), Mattson (1984), and Durham (1985), give strong support to the idea that the Greater Antilles formed on a fracture zone in the eastern Pacific Ocean during the Cretaceous. According to this thesis, as North and South America moved westward, the Greater Antilles moved into the gap between the continents. Subsequent movement in the same direction supposedly caused the Greater Antilles to shift into the Caribbean and then drift northeastward until they collided with the Bahamas Platform. In the meantime, Central America was supposed to have formed as an island arc in the eastern Pacific and then to have moved eastward to occupy its present position.

The primary difficulty with the foregoing consensus is that it is based largely on conjecture since many of the traditional parameters used to establish plate motions (magnetic lineations, extensive fracture zones, paleomagnetic positions, and others) are lacking or are not applicable (Smith, 1985). There are two other interpretations which suggest that the Greater Antilles and Caribbean plate formed much closer to home as North and South America separated. Anderson and Schmidt (1983) believe that the Greater Antilles split off from northern South America. Donnelly (1985) suggested that parts of the Greater Antilles have never been far from their present location. He also noted that biogeographically significant Tertiary tectonics in Hispaniola, Cuba, and much of Central America have been dominated more by local vertical tectonics than by putative horizontal movements. Local geological evidence for the Antilles indicates that their subaerial emergence has been comparatively recent. According to Malfait and Dinkelman (1972), post-Eocene movements led to the uplift of the present islands of Cuba, Hispaniola, and Puerto Rico. It was apparently during the Miocene that Cuba became completely emergent (Rigassi, 1963; Mattson, 1984). Jamaica has been reported as being totally submerged during the Miocene (Robinson and Lewis, 1971) or as beginning an uplift in the mid-Miocene (Arden, 1975). On Hispaniola, the emergence and erosion of the Central Cordillera followed a late Eocene event (Biju-Duval et al., 1982). For Puerto Rico, Monroe (1980) described the uplift of the central core of the island as taking place in middle Miocene time.

It was Matthew (1915, 1918) who presented the first cogent argument in favor of the colonization of the Antilles by over-water dispersal. He pointed out that the vertebrate fauna, fossil and recent, represents only a few selections from the continental faunas of either North or South America and that it is an unbalanced assemblage that had apparently arrived at diverse times. His theory was strongly reinforced, primarily on the basis of the mammalian fauna, by Darlington (1938, 1957) and by Simpson (1956). MacFadden (1980) felt that the Antillean insectivores *Nesophontes* and *Solenodon* might be Mesozoic relicts and could have arrived in accordance with Rosen's theory. But Simpson (1956) had considered these forms to be early Tertiary in origin and no significant new evidence was presented by Mac-Fadden.

Of all the animal groups present on the Antilles, it is perhaps the freshwater fishes that offer the best clue to the origin of the Antillean biota. While almost all other groups allow, at some stage in their life history, the possibility of transport by rafting or by hurricanes or by birds, fishes must remain in the water. The primary freshwater fishes of the world belong to a huge complex of 57 families that are almost completely intolerant of saltwater. When Myers (1938) published his account of freshwater fishes and West Indian zoogeography, he noted that a striking feature of the fauna was the complete absence of primary freshwater fishes. An analysis of the phylogenetic relationships of the Antillean secondary freshwater fishes has been published by Briggs (1984b).

In the northern Antilles from Cuba to Puerto Rico the progressive decrease in the diversity of the secondary freshwater fish fauna is remarkable (Table 1). Cuba is the only island that possesses endemic genera (2), but there is one endemic Antillean poeciliid genus (*Limia*) that is found on Hispaniola, Cuba, Jamaica, and Grand Cayman. *Limia* is most speciose on Hispaniola (17 species); it probably originally developed there and then spread to the other islands. There is also one endemic Antillean cyprinodontid genus (*Cubanichthys*) found on Cuba and Jamaica. These genera, possibly dating back to the Miocene or Oligocene, probably represent the earliest island colonizations.

Jamaica, the Bahamas, and Grand Cayman each have endemic species but no such genera. Such colonizations (to these and other islands) probably occurred later, perhaps in the Pliocene or Pleistocene. Finally, the Antillean species that are cur-

TABLE 1

Diversity and endemism of secondary freshwater fishes in the southern Antilles[a,b]

Family	Cuba		Hispaniola		Jamaica		Bahamas		Grand Cayman	
	gen.	sp.	gen.	sp.	gen.	sp.	gen.	sp.	gen.	sp.
Poeciliidae	4 (2)	16 (14)	3	25 (24)	2	7 (5)	1	2 (1)	2	3 (2)
Cyprinodontidae	4	7 (4)	2	3 (1)	2	2 (1)	2	3 (1)		
Cichlidae	1	2 (2)	1	1 (1)						
Synbranchidae	1	1								
Lepisosteidae	1	1 (1)								
Total	11 (2)	27 (21)	6	29 (26)	4	9 (6)	3	5 (2)	2	3 (2)

[a] Number of endemic forms given in parentheses.
[b] No species of these fishes are native to Puerto Rico.
gen.: genera; sp.: species.

rently shared with mainland localities such as *Cyprinodon variegatus, Rivulus marmoratus, Fundulus grandis,* and *Ophisternon aenigmaticum* probably represent Pleistocene or recent invasions. Thus, the phylogenetic relationships of the Antillean freshwater fish fauna indicate that a series of invasions by these saltwater tolerant forms took place over a period of time from about the mid-Tertiary to the present. The initial colonizations took place on Cuba and Hispaniola, probably soon after these islands emerged from the sea and developed freshwater stream systems.

The relationships of the West Indian herpetofauna have been summarized by Duellman (1979). There is a total of 505 species in 57 genera; 476 species and 18 of the genera are endemic. A phylogenetic and biogeographic analysis of the xenodontines, a subfamily group of the snake family Colubridae, has been published by Cadle (1985). He concluded that the most likely hypothesis to explain the origin of the West Indian xenodontine fauna is a late Tertiary dispersal from South America and subsequent radiation within the Greater and Lesser Antilles. Further, his observations suggested over-water dispersal as a plausible explanation for the occurrence of these snakes in the West Indies.

Trueb and Tyler (1974), who studied three hylid frog genera, suggested that five invasions of the Greater Antilles by separate hylid stocks took place, probably by means of rafting from South America. The West Indian bird fauna consists of about 300 species. Thirty one of the genera and 125 of the species are endemic; there is one endemic family (Brown and Gibson, 1983). Lack (1976) found that, of the 66 species of resident land birds of Jamaica, 25 also occur on the nearest mainland and at least 20 others have their closest relatives there. The fact that six genera are endemic to Jamaica is interesting in view of the information that Jamaica has possibly been emergent for only about 15 Ma (Robinson and Lewis, 1971).

Pregill (1981) wrote an appraisal of the biogeography of the West Indian terrestrial vertebrates. He observed that the fossil record for North and South America indicates that the mid-Tertiary is the probable time when most of the living genera appeared in the West Indies and that colonization has continued ever since. Pregill also observed that the distributional facts are consistent with the hypothesis of an insular biota that has evolved through time from organisms that have come over water from different places.

It may be observed that the historical terrestrial vertebrate pattern is similar to that presented by the secondary freshwater fishes. There are endemic genera in the Greater Antilles that probably arrived in the mid-Tertiary and the relationships of the species indicate that a series of invasions, mainly from South and Central America, must have taken place over a long period of time.

There have been only a few published speculations about the relationships of Antillean insects. Analyses at the species level of the Trichoptera and Odonata faunas led Flint (1978) to conclude that their distribution could be explained by the vicariance model of Rosen (1976). A similar, late Mesozoic island drift hypothesis was invoked by Shields and Dvorak (1979) to explain the relationships of certain butterfly species. On the other hand, Erwin (1979), in discussing the carabid beetles, observed that there was much to argue the case for dispersal to the islands by air or wood drift and very little to argue the case for vicariance.

Although the flora of the West Indies, especially that of the Greater Antilles, is distinguished by many endemic genera and species there is a high diversity and a representation of most mainland groups. Carlquist (1974) observed that the Antillean wet forest is relatively harmonic in that it contains many of the same species seen in mainland wet forests. A primitive cycad (*Microcycas calocoma*) considered to be a Mesozoic relict, occurs on Cuba (Carlquist, 1965). Otherwise, the floral relationships appear to be more recent. Raven and Axelrod (1974) indicated that the notion of a Paleogene arrival of some South American taxa in the West Indies would be consistent with the prominence of such groups as the Cactaceae, Gesneriaceae, and Canellaceae.

In general, it may be noted that the Antilles possess a disharmonic or unbalanced fauna while the flora is relatively harmonic. As Darlington (1957), who analyzed the vertebrate fauna as whole, observed, the island species represent those mainland groups that are most likely to disperse across saltwater. The situation is quite analogous to that of Madagascar which has been separated from Africa since the mid-Jurassic (p. 161). Each location has a limited small mammal fauna that has evidently accumulated over a considerable time, the other terrestrial vertebrates also represent a depauperate sampling of the mainland fauna, and there are no primary freshwater fishes.

The land bridge hypothesis has not been seriously considered in recent years. There is no geological evidence to support it and if one builds a bridge to permit a few species to cross there is always the problem of having to destroy it quickly before too many other species get across. The migrating island idea is objectionable because the Antilles apparently did not finally emerge from the sea until post-Eocene times and, with the possible exception of one cycad species in Cuba, there are no Mesozoic relicts to be found. Population by fortuitous or waif dispersal, probably beginning in the mid-Tertiary and continuing to the present, is indicated by the phylogenetic relationships of the island species. This leaves little doubt that the Antilles should be regarded as oceanic islands that have gradually accumulated their terrestrial and freshwater biota by means of overseas transport.

CENTRAL AMERICA

When the Isthmus of Panama was finally completed about 3 – 5 Ma, the subsequent dispersal events, at least as far as terrestrial animals are concerned, are reasonably well known (Webb, 1978; Simpson, 1980; Marshall et al., 1982). However, for the Cretaceous and the early Tertiary, information about the biotic and geological events is only fragmentary. So far, most of the speculation about the history of Central America has been based on geophysical evidence, but a viable hypothesis must also be consistent with the biological data.

As a group, the mammals have the best fossil record and can therefore provide more information about interamerican relationships than any other animal group. An important event was the discovery of latest Cretaceous marsupials from near Lake Titicaca in Peru (Sigé, 1972; Archibald and Clemens, 1984). Didelphid and possibly pediomyid marsupials, both common in the latest Cretaceous of North

America, have been found at the Titicaca site. A possible condylarth has also been reported. The discovery of contemporaneous fossils in South America, when considered in the light of the major marsupial evolutionary radiation that took place on that continent, could be persuasive toward a South American origin (Archer, 1984). Condylarths were unknown previous to the Paleocene (Colbert, 1980). It may be that, they too, migrated northward from South America.

The early hoofed mammals, the notoungulates, underwent an enormous diversification in the early Tertiary of South America, but a few species have also shown up in the late Paleocene of Asia (Savage and Russell, 1983). In a like manner, the edentates began to blossom in the Eocene of South America, but a single fossil has been identified from the late Paleocene of China (Ding, 1979). Another early Tertiary development was probably the migration of a primitive primate from North to South America to begin the evolution of the New World monkeys (Simpson, 1980). Also, the evidence seems to indicate that the caviomorph rodents reached South America from the north in late Eocene or early Oligocene times by rafting (Colbert, 1980).

By the Miocene, the presence of mammalian fossils of North American affinity in the Panama Canal area (the Cucaracha Formation) indicates that the isthmian region had considerable contact with North America (Savage and Russell, 1983). In the late Miocene, North America received the first mammalian immigrants from South America since the late Paleocene (Webb, 1985a). These were two genera representing two different families of ground sloths. Presumably they crossed considerable water gaps for they came alone and were not succeeded by other South American genera for another 4 million years or so. At about the same time, a genus of the raccoon family showed up in South America (Webb, 1985b).

It was toward the end of the Pliocene that the first great group of South American mammals appeared in North America. The most diverse were the edentates consisting of a sloth and three armadilloid genera. Two hystricognath rodents appeared, a porcupine and a large amphibious capybara (Webb, 1985a). Finally, by the early Pleistocene, the great interchange was underway. Fifteen families of North American land mammals arrived in South America and the largest group of South American mammals had dispersed northward. It is now apparent that only during the past 3 million years has Central America provided an effective migration corridor for mammals. North-south migrations during the Tertiary were comparatively few and difficult.

In the herpetofauna, if has been noted that late Cretaceous hadrosaurs, common in the northern hemisphere, have been recorded from several localities in the Upper Cretaceous of Argentina (Estes and Baez, 1985). The fossil record of lizards has become better known in recent years (Estes, 1983). A primitive iguanian group probably arose in the middle to late Jurassic of South America – Africa – Madagascar when those lands were joined or nearly so. This was followed by a late Cretaceous or early Cenozoic dispersal to North America. The teiid and anguid lizards probably evolved in North America during the Cretaceous then migrated to South America shortly afterward. The skinks (Scincidae) apparently arrived in North America via the Bering Land Bridge in the Cretaceous then, by the Paleocene, made their way to South America.

In regard to snakes, the genus *Coniophis,* of uncertain family status, is known from the late Cretaceous of Bolivia and from the late Cretaceous to the Eocene of North America (Estes and Baez, 1985). The family Elapidae, represented by the living genus *Micrurus* in Central and South America, must have come from North America and was probably established in its present area by at least the Miocene (Savage, 1982). The South American fer-de-lance (*Bothrops*) and the bushmaster (*Lachesis*) are members of the poisonous family Viperidae which probably made its way from Asia across Beringia in the Eocene and reached South America by Oligocene to Miocene times (Brattstrom, 1964).

The amphibian data is interesting. The salamander family Plethodontidae evolved in North America and then dispersed south along the upland areas of Central and northern South America (Wake and Lynch, 1976). The South American species belong to the genus *Bolitoglossa* and are derived from a Central American radiation of that group. Since two of the species groups are endemic to South America, it has been suggested that the genus entered South America prior to the establishment of the isthmian link (Vanzolini and Heyer, 1985). The frog family Bufonidae probably originated in South America (Blair, 1972). The genus *Bufo* arrived in Central America in the early Oligocene and underwent a modest radiation. Descendents of that radiation reached North America in the middle to late Oligocene (Maxson, 1984; Vanzolini and Heyer, 1985).

The New World tree frogs of the family Hylidae originated in South America and reached Central America in the early Tertiary. In Central America they underwent a large radiation which gave rise to the North American forms beginning in Eocene/Oligocene times (Maxson and Wilson, 1975; Vanzolini and Heyer, 1985). The frog family Leptodactylidae also originated in South America (Heyer, 1975). Extensive leptodactylid radiations have occurred in both the West Indies and Central America. The size of the radiations indicate ancient distributional events. The North American members of the family were probably derived from both radiations (Vanzolini and Heyer, 1985).

As with the foregoing frog families, the Microhylidae evolution in the New World probably began in South America. The North American taxa are most directly related to those of Central America. The migration to North America probably took place well before the formation of the isthmian link (Vanzolini and Heyer, 1985). The frog family Ranidae is represented by only one genus, *Rana*, in the New World. The other genera occur in Asia and Africa. It seems clear that *Rana* reached North America via Beringia then migrated to Central America where a modest radiation took place. One member of that radiation has entered South America, probably after the formation of the isthmus (Vanzolini and Heyer, 1985). Still earlier amphibian relationships are indicated by the occurrence in Argentina of the early Jurassic genus *Vieraella* (Estes and Reig, 1973). That genus belongs to the family Ascaphidae which is represented by a living genus (*Ascaphus*) in northwestern North America. A late Jurassic fossil from South America (*Notobatrachus*) seems to be related to *Ascaphus* and *Leiopelma*. The latter is a living genus in New Zealand.

A detailed analysis of the entire Central American herpetofauna and its history was presented by Savage (1982). He recognized three major dispersal events: (1) a movement of southern forms into Central America in the late Mesozoic and/or early Tertiary; (2) a dispersal of northern stocks into Central America prior to the Eocene; and (3) a second dispersal from South America when the isthmian link was completed. The first two dispersals were considered to have been followed by vicariant events that cut off gene flow and allowed considerable differentiation to take place in Central America. The presence of 32 endemic genera (out of a total of 169) indicate a major evolutionary radiation that must have gotten underway in the late Cretaceous or very early Tertiary. Vanzolini and Heyer (1985) pointed out that, in almost all cases, the herpetofaunal interchanges between North and South America were not accomplished directly but through the medium of Central American radiations.

In Central America, the primary (ostariophysan) freshwater fishes are rather poorly represented (Miller, 1982). It has been suggested that the species of this group were not able to invade Central America until a complete isthmian connection was formed in the late Pliocene or early Pleistocene (Myers, 1966). However, the killifish family Poeciliidae, a secondary freshwater group, has undergone a significant evolutionary radiation in Central America as is indicated by a high level of endemism including taxa above the generic level (Rosen and Bailey, 1963). Bussing (1976, 1985), in discussing the distribution of freshwater fishes, reviewed the earlier geological and geophysical literature. He concluded that substantial portions of the Central American archipelago were in place by Upper Cretaceous/Paleocene times, and that these portions remained in place, with some changes in configuration, throughout the Tertiary (Fig. 5). He was able to divide the fishes into an Old Southern Element that reached Central America from South America in the late Cretaceous or early Tertiary and a New Southern Element that began to reach Central America sometime in the Pliocene.

It has been noted (p. 25) that the dipterid family Drosophilidae apparently originated in the Old World tropics and, by Eocene time, had undergone a series of radiations (Throckmorton, 1975). The dispersal of the family involved multiple crossings of the Bering Land Bridge followed by a series of invasions southward to

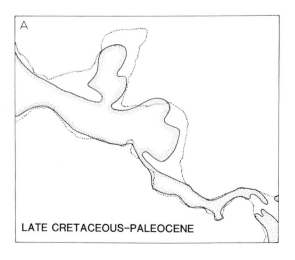

A

LATE CRETACEOUS-PALEOCENE

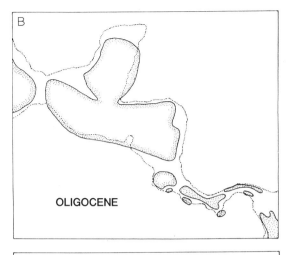

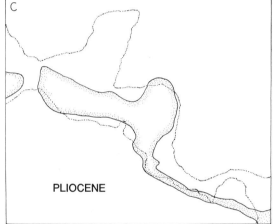

Fig. 5. Approximate extent of terrestrial habitat in Central America at three stages of its development. Dotted line indicates the modern configuration. After Bussing (1976).

the neotropics. From South America a number of northward reinvasions of the temperate zone occurred. Edmunds (1975), in discussing the distribution of the mayfly (Ephemeroptera) families Oligoneuriidae and Tricorythidae, observed that the North American members were derived from South American lineages.

Raven and Axelrod (1974), in their comprehensive work on angiosperm biogeography, presented a detailed analysis of North American – South American relationships. Those families emigrating from South America northward in the early Tertiary include the Cactaceae, Liliaceae-Allieae, Loasaceae, Martyniaceae, Nyctaginaceae, Tecophilaeaceae, and Zygophyllaceae. Early migrators in the opposite direction were Boraginaceae, Clethraceae, Gentianaceae, Hydrophyllaceae, Scrophulariaceae, and others. By the late Tertiary, the intercontinental migratory movements become much more pronounced. Raven and Axelrod suggested that at least the great majority of temperate North American plants did not appear south of the Isthmus of Tehuantepec until the Upper Miocene, whereas South American plants became established in the tropical portions of the North American region from at least Eocene time onward. They concluded that the tropical biota of South America has progressively taken over areas of appropriate climate in North America (including Central America and the West Indies), just as the tropical biota of Asia have taken over the lowlands of New Guinea and neighboring islands.

The interchange of various plants in particular habitats was discussed by Simpson and Neff (1985). They noted that a few lowland tropical elements, such as members of the Aristolochiaceae and the Vitaceae, moved from North America into South America. But that many species now found in Panama and countries to the north belong to genera that are "centered" in South America. In the montane habitat, there are some widespread Holarctic elements that arrived in South America by mountain hopping. Several important plant groups have disjunct distributions between the deserts of North and South America. Examples are genera in the Cactaceae, Celastraceae, Fabaceae, and Verbenaceae. The creosote bush, *Larrea,* evidently originated in South America and migrated northward while other genera, such as *Agave, Yucca, Parthenium,* and *Zaluzania,* apparently traveled in the opposite direction. The Mediterranean scrub vegetation of the two continents also demonstrates some relationships. Genera of presumed northern origin that are now found in the Chilean matorral include *Myrica, Salix, Alnus, Berberis,* and *Ribes.* Almost all such relationships are probably attributable to long distance dispersal.

Although the foregoing data are fragmentary, they do provide valuable information about North American – South American relationships during the late Mesozoic and Tertiary. Since the two continents were well separated by the late Jurassic (Map 2), an island archipelago, which served as a filter bridge, must have been in place by the latest Cretaceous. Otherwise, it would be difficult to account for the relationships shown by such mammalian groups as the didelphid marsupials, condylarths, notoungulates, edentates, and reptiles (including certain dinosaurs, lizards, and snakes). Furthermore, there are reasonable indications that intercontinental traffic also took place during the early Tertiary from the Paleocene to the Miocene. Groups involved were apparently the omomyid primates, caviomorph rodents, lizards, elapid snakes, viperid snakes, some frog family groups, some insects, and several plant families. These are indications that, once established, the archipelago filter bridge remained more or less available during the early Tertiary. The secondary freshwater fish family Poeciliidae and the herpetofauna in general have undergone a significant evolutionary radiation in Central America providing more evidence that a spacious archipelago must have been available from the late Cretaceous until the isthmus was established.

It may been seen that, at our present state of knowledge, the geophysical scenarios provided by many recent authors are not consistent with the biological data. The proto-Antilles concept of most of these authors requires that this original archipelago be moved out of the gap between the Americas in about the late Cretaceous (110 – 60 Ma) and not be replaced by a new Central American archipelago until the early Miocene. The contrasting hypotheses of Anderson and Schmidt (1983) and Donnelly (1985) better suit the biological data. They envision the formation of Central America approximately in situ as North and South America drifted apart. This would provide the necessary filter bridge for terrestrial animals and plants and permit the evolution of the freshwater poeciliid fishes and the endemic herpetofauna in the Central American area.

OTHER BARRIERS

Hallam (1981) referred to the formation of the interior seaway in North America during the mid-Cretaceous as the most significant biogeographical event in the continents of the northern hemisphere because it created two separate land masses. However, as Cox (1974) showed in his study of fossil vertebrate relationships, the northern part of the Pangaean land mass (Map 1) was probably first split by the intrusion of the Turgai Sea in the mid-Jurassic producing a continental Euramerica and a separate Asia (Map 2). By the early Cretaceous, Euramerica had lost its connection to Africa with the formation of a complete Tethys Sea (Map 3). So, by the mid-Cretaceous, when the Mid-Continental Seaway split off the western part of Euramerica, it produced not two but three northern landmasses, Westamerica, Euramerica, and Asia (Map 4). Soon after, in the late Cretaceous, the Beringia connection took place reducing the three areas to two, a Euramerica and an Asiamerica (Map 5). The disappearance of the Mid-Continental Seaway at the end of the Cretaceous allowed the formation of a single, northern land mass interrupted only by the Turgai Sea (Map 6).

The single, northern continent that emerged at the end of the Cretaceous persisted until the early Eocene when, as the biological data indicated, the connection between Europe and North America was severed producing a European continent and an Asiamerica (Map 7). This situation lasted until the Oligocene when the Turgai Sea regressed allowing Europe to finally become united to Asia (Hallam, 1981). So, by Miocene times there was again a single northern land mass with the only circumpolar break being that located between Europe and North America (Map 8). Later, this single terrestrial continent became two during the relatively brief intervals when the Bering Land Bridge was inundated and as it is now.

Our knowledge about the existence and location of the two gigantic epicontinental seas (Mid-Continental and Turgai) comes almost entirely from studies of fossil remains of the biota that was affected by them. The primary cause for the appearance and disappearance of the seas was eustasy or changes in the sea level of the world ocean (Hallam, 1981). There has been a tendency for earth scientists to overlook the presence of these shallow seas so they are usually not depicted on maps showing the historical positions of the continents. Yet, as the paleontological record shows, the biogeographical effects were very important.

The drying up of the Turgai Sea in early Oligocene, after having served as a virtually continuous sea barrier between Asia and Europe since the Jurassic, had an immediate influence on the European mammalian fauna. This major faunal change has been called the "Grande Coupure" and its effects have been reviewed by McKenna (1983b). Seventeen western European terrestrial mammalian genera became extinct. At the same time, 20 new genera appeared, most of which seem to have originated elsewhere. McKenna considered the new arrivals to have come from Asia and southeastern Europe.

For the marine animals of the shelf, the two epicontinental seas provided a productive, warm-water environment. During the Cretaceous, their presence was outlined by distinct provinces of bivalve molluscs (Kauffman, 1973) and by the distribution patterns of the late Cretaceous Ammonoidea (Matsumoto, 1973). At the same time, the seas constituted important barriers to the dispersal of terrestrial and freshwater animals and plants. The evidence from the land vertebrates of the Upper Cretaceous is consistent with the idea of a Euramerica, bordered by the Turgai Sea to the east and the Mid-Continental Seaway to the west, and an Asiamerica which was continuous across the Bering Strait. For example, the tyrannosaurs and protoceratopsids occurred only in Asiamerica while the iguanodonts and primitive ankylosaurs survived in Euramerica (Cos, 1974; Hallam, 1981). Palynological research has led to the recognition of separate floral provinces for the same continental areas (Muller, 1970; Hickey, 1984; Tschudy, 1984).

Chapter 4

THE INDO-AUSTRALIAN CONNECTION

> My next paper . . . was on the birds of the chain of islands extending from Lombok to the great island of Timor. I gave a list of one hundred and eighty-six species of birds, of which twenty-nine were altogether new; but the special importance of the paper was that it enabled me to mark out precisely the boundary line between the Indian and Australian zoological regions . . .
>
> Alfred Russel Wallace, *My Life,* 1905

Thanks to Alfred Russel Wallace's field work in the Malay Archipelago and his subsequent publications, considerable attention became focussed on that area during the latter half of the 19th century. He was the first to call attention to the dramatic changes that took place in the fauna as one traveled along the island chain. In 1863, Wallace presented a paper to the Royal Geographical Society in which he drew a line starting east of the Philippines and extending south to separate Borneo from Celebes and Bali from Lombok. By 1880, his publications, including extensive works such as *The Malay Archipelago* (1869), *The Geographical Distribution of Animals* (1876), and *Island Life* (1880), had drawn the attention of naturalists all over the world. Interest in the Malay Archipelago has continued to the present day. There are a number of modern works available including a book devoted to Wallace's Line (Whitmore, 1981a).

When Wallace's work began to be published, other investigators explored the Malay Archipelago and drew alternative lines based on the distribution of various animal groups. So a number of choices became available, including two suggested by Wallace (Fig. 6). Most modern biogeographers have preferred his 1863 line. The sharp contrast in the fauna from one side of the Line to the other is not reflected to the same extent by the plant life. In general, plants appear to be better at overseas dispersal as demonstrated by the relatively rich and harmonic floras found on such isolated islands as the West Indies and Madagascar.

Although different groups of animals demonstrate the effect of Wallace's Line in different ways, depending mainly on their ability to get across saltwater barriers, the overall effect is one of a very sharp change across only a few miles of separation, a condition that is not found in any other part of the world. For many years, this mystery was explained by the fact that the islands to the west of the Line were situated on the Sunda continental shelf. When the eustatic sea level was lowered during the ice ages, those islands became continuous with the Asian mainland. This explained the oriental nature of the fauna of the western islands but did not explain why the entire island chain, over eons of time, did not develop a more homogeneous fauna. The problem was not solved until it was realized that the tectonic plate bearing Australia – New Guinea had moved north with its peculiar biota in the relatively recent past.

The Mesozoic and Cenozoic movements of the Australian plate have been illustrated by Audley-Charles et al. (1981). They show the plate moving northward at about 40 Ma and approaching close to Indo-Malaya about 20 Ma. It apparently moved into about its present position in Miocene/Pliocene times. It was then that the two different biotas made contact. The contribution of the Australian continental crust to the Malay Archipelago has been mapped (Audley-Charles, 1981) (Fig. 7).

The vertebrate fauna of the Archipelago has been reviewed by Cranbrook (1981) and Keast (1983). Of all the vertebrate groups, it is the primary freshwater fishes that most clearly demarcate Wallace's Line. The rich primary fauna of southeastern Asia, consisting of such families as Cyprinidae, Gastromyzontidae, Homalopteridae, Cobitidae, Siluridae, Bagridae, etc., that are represented in Borneo have been completely unable to reach Celebes only 65 miles across the Makassar Strait. The snakehead, *Ophicephalus striatus*, and the climbing perch, *Anabes testudinius*, do occur in Celebes but these are food fishes with accessory air-breathing organs and are presumed to have been carried there by man (Cranbrook, 1981).

To the north of Borneo, a few primary freshwater fishes have invaded the Philippines but this area lies to the west of Wallace's Line and is best considered an isolated outpost of the Oriental Region. To the south, the family Cyprinidae is represented on Lombok by two genera, *Puntius* and *Rasbora*, each with a single species. *Rasbora* extends eastward to the next island (Sumbawa) but the final limit for primary freshwater fishes is reached at Sape Strait just east of Sumbawa. The freshwater fish fauna of the islands farther east, and of New Guinea and Australia, has been entirely derived from the surrounding marine environment. As noted (p. 73), the Australian lungfish, *Neoceratodus*, and the osteoglossid, *Scleropages*, both

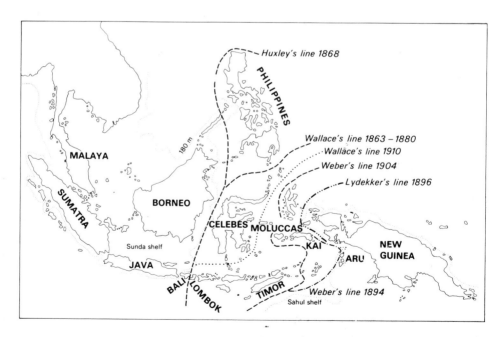

Fig. 6. Various lines that have been proposed to mark the area of greatest faunal change. Wallace's (1863–1880) line has continued to receive the most support. After George (1981).

have marine ancestors so probably did not reach Australia via freshwater pathways.

Among the amphibians, there are five species of caecilians on Borneo but none on Celebes. There are almost 100 species of frogs on Borneo but only 23 species on Celebes (Cranbrook, 1981). Most of the latter are endemics; of these, four show western affinities and two eastern. To the south, along the Lesser Sunda chain, there is a progressive depletion of the western fauna and a corresponding increase in eastern relationships. In this group, the Lombok Strait does not comprise an important barrier. Instead, there is only a gradual decrease in the oriental fauna as one proceeds eastward (Table 2).

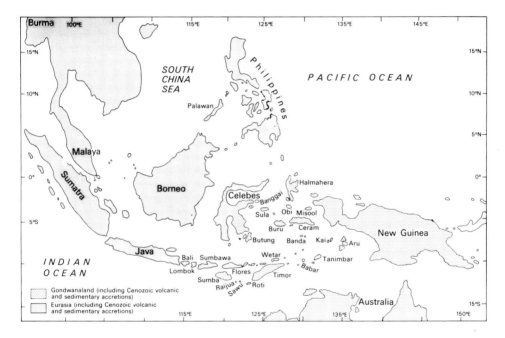

Fig. 7. The modern Malay Archipelago showing the main contributions of the Australian Plate (stippled). After Audley-Charles (1981).

TABLE 2

Percentage representation of vertebrate groups of oriental origin as they extend eastward on the islands of the Lesser Sunda chain (after Cranbrook, 1981)

	Bali	Lombok	Sumbawa	Flores
Amphibians	92	78	75	70
Reptiles	94	85	87	78
Birds	87	73	68	63

The rich oriental reptilian fauna also becomes gradually attenuated as it extends eastward along the Malay Archipelago. There appears to be a significant drop off from Borneo to Celebes. For example, the number of snake species recorded from Borneo is 166 but this drops to only 61 on Celebes. But along the Lesser Sunda chain, there is a gradual decrease in diversity comparable to that noted for the amphibians (Table 2). Oriental groups that have reached continental Australia include the lizard families Varanidae, Scincidae, and Agamidae; the turtle families Carrettochelyidae and Trionychidae; and the snake families Elapidae, Colubridae, Acrochordidae, Uropeltidae, Typhlopidae, and Boidae (Tyler, 1979).

The diversity among bird species also declines in an eastward direction. On Borneo there are about 396 species and about 220 on Celebes (Cranbrook, 1981). But, the relationships of the Celebes birds are primarily oriental. Along the Lesser Sunda chain the eastward decrease is more gradual (Table 2). Keast (1983) provided a table that gives the composition of the fauna for all of the main islands between the Malay Peninsula and New Guinea. Many oriental families extend well along the archipelago and some of them have invaded continental Australia. On the other hand, many of the endemic Australian families do not extend beyond the continental limits.

It seems apparent that the Australian birds, unlike the other vertebrate groups, have occasionally given rise to groups that have successfully spread to other parts of the world. It has been suggested that the parrots and the pigeons are examples of such groups (p. 69). Sibley and Ahlquist (1986) have determined that the corvid birds (crows, ravens, magpies, jays, and their relatives) first evolved in Australia then succeeded in invading southeast Asia about 30 Ma. The migration route was probably via New Guinea and then westward along the archipelago. In the Oriental Region, they underwent a second evolutionary radiation before spreading to the rest of the world. The same path may have been followed by the pigeons and the parrots, although those groups are still most diverse in the southern hemisphere.

The mammalian fauna of the Malay Peninsula, Sumatra, Java, and Borneo is exclusively oriental. As with the other terrestrial vertebrates, there is a decrease toward the east. There are about 200 species on Borneo and some 108 on Celebes (Cranbook, 1981). On Celebes, there are several widespread Indo-Australian bat species but, among the non-flying mammals, there are only two endemic species of the marsupial genus *Phalanger*. All the others are clearly of Asiatic derivation. It has been noted (p. 71), that placental mammals of the rodent family Muridae are abundant in Australia. They apparently got there by migrating across the archipelago in the early Pliocene. The oldest rodent fossils found in Australia are four to five Ma old. Invasions by the genus *Rattus* are more recent, probably taking place in the Pleistocene.

Australia has been invaded from the Oriental Region by many invertebrate groups. Ross (1967) noted the presence of a group of modern caddisfly genera that had come from southeast Asia. The origins of the Australian waterbug (Hemiptera) families are probably the same (Lansbury, 1981). Williams (1981) noted that, in general, the Australian aquatic insect fauna consisted of a large southern element of primitive forms with austral affinities and a younger, northern element with oriental relationships. Among the terrestrial oligochaete worms, Jamieson (1981) identified a small endemic component that was the result of post-Miocene invasions from the Oriental Region. In the terrestrial insects of Australia, there are many modern genera of southeast Asian origin (Mackerras, 1970).

The distribution of various groups of plants along the Malay Archipelago has been considered most recently by van Steenis (1979), Dransfield (1981), and Whitmore (1981b). The work by Whitmore called attention to certain Laurasian families that had penetrated the archipelago from the west and compared their distributions to other groups that had originated in Gondwanaland and entered from the east. For example, there is the Dipterocarpaceae, which ranges from Africa and India through China but has a center of diversity in Sumatra, Malaya, and Borneo. In these large islands there are 10 genera and more than 280 species. Celebes has but two genera and 45 species and eastward along the Lesser Sunda chain the diversity decreases rapidly. Another example is the primitive family Magnoliaceae. Although widespread, it is strongly concentrated in East Asia and progressively diminishes as it reaches New Guinea.

Examples of Gondwana groups provided by Whitmore (1981b) are *Phyllocladus* (Podocarpaceae), *Agathis* and *Arucaria* (Arucariaceae), *Styphelia* (Epacridaceae), and the family Winteraceae. These are groups that were supposedly carried north on the Australian plate. Upon reaching the Malay Archipelago, they were evidently able to extend to some degree into the Oriental Region. That this actually happened seems a reasonable assumption. However, one must keep in mind that the Arucariaceae is an ancient conifer group that was once widespread in the northern hemisphere (Florin, 1963). Also, the Podocarpaceae extends well north in the Pacific to the Ryu Kyu Islands and Japan. In the New World it extends north to Mexico and the West Indies. Such northward extensions could be due to factors other than continental drift.

The primitive angiosperm family Winteraceae is considered to be a southern counterpart to the Magnoliaceae. Most of the genera in the former are restricted to the Australian – New Zealand area. One genus, *Drimys*, ranges westward into Borneo and the Philippines. There are also four species in southern South America. The family Proteaceae has evolved primarily in the Australian area (Johnson and Briggs, 1975). Four genera extend into the Malay Archipelago. But, one genus, *Heliciopsis,* is endemic to southeast Asia and the archipelago west of Wallace's Line. Johnson and Briggs suggested that it could have reached southeast Asia by passage aboard India as it moved northward. The genus is not found in India today but proteaceous pollen has been reported from the Eocene of that country.

The distribution of the palm family (Arecaceae) in the Malay Archipelago has been analysed by Dransfield (1981). He followed Moore (1973) in the assumption that the palms had their origin in west Gondwanaland (South America) and have subsequently radiated from that area. The palm floras of the east and west parts of the archipelago were found to be, in many ways, markedly different. Some of the patterns suggested invasions of the island chain from two directions, from New Guinea and from southeast Asia, following the Miocene arrival of the Australian plate.

The family Fagaceae is well developed in southeast Asia (Whitmore, 1981b). Several of the genera extend out into the Malay Archipelago. *Trigonobalanus*

50

reaches Celebes and *Castanopsis* and *Lithocarpas* both range all the way to New Guinea. In the latter island, they overlap with *Nothofagus* which is represented on New Guinea by 16 species. This genus is placed in the subfamily Fagoideae along with the northern beech *Fagus*. The latter is found in southern China and is widespread elsewhere in the northern hemisphere. It has been noted (p. 65) that *Nothofagus* is broadly distributed in the southern hemisphere. So, one may observe, on one hand, that *Nothofagus* was moved northward by riding on the Australian plate but, on the other hand, the distribution of the various genera appear to indicate that the Fagaceae originated in southeast Asia. If the family originated there, then the Fagoideae must have also come from that area. So this must have been the region in which *Fagus* and *Nothofagus* originally parted company, one expanding northward and the other to the south.

Assuming that the angiosperms as a group originated in southeast Asia (Briggs, 1984a), there must have been a route whereby some of the primitive forms such as the Winteraceae, *Nothofagus* and perhaps the early palms, reached the continents of the southern hemisphere. A connecting land mass in the mid-Cretaceous or earlier was postulated by van Steenis (1979) but this was probably not the case since there are no indications of early animal life that would have taken advantage of such a connection. The primitive fauna that accumulated in Australia and New Zealand

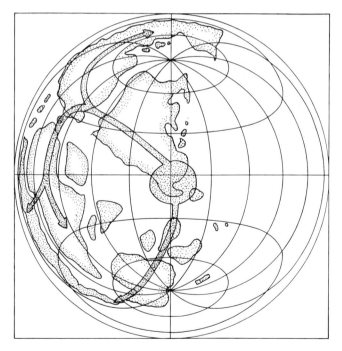

Fig. 8. Hypothetical dispersal of primitive angiosperms and other plantlife from southeast Asia in the early Cretaceous. Australia – New Guinea (shown here divided into three islands by epicontinental seas) was, at this time, situated closer to the Malay Peninsula area than at any other time until the mid-Tertiary. Map redrawn from Audley-Charles et al. (1981) and other sources.

evidently came from South America via Antarctica. What was needed in the early Cretaceous was a filter route negotiable by plants with good dispersal ability but too difficult for most animal species.

A series of plate tectonic maps of the western Pacific, based on geophysical data, were published by Audley-Charles et al. (1981). They indicate that, at about 120 Ma, the northern part of the Australia – New Guinea area and the southern part of the Malay Peninsula were in closer proximity than they were during the succeeding 60 Ma. The oceanic gap illustrated for 120 Ma (Fig. 8) is not so great as to completely preclude passage by storm driven propagules, especially if there happened to be an island or two along the route.

The foregoing scenario would permit an early southward dispersal of some of the primitive angiosperms (and probably conifers and hepatophytes as well). Then, following a southern hemisphere radiation, during about the last half of the Cretaceous and the first half of the Tertiary, the Australian plate could have transported the descendents of the early Cretaceous dispersal back toward their original home. This could account for the dual phylogenetic and dispersal patterns that are evident in the modern Malay Archipelago.

In an article that dealt with Wallace's Line as a zoogeographic barrier (Briggs, 1974), it was noted that a number of species of oriental frogs, reptiles, birds, and mammals had managed to get across the barrier and, in many cases, had succeeded in extending their ranges into Australia. In contrast, very few Australian species, mainly birds, had been able to get past the Line in the opposite direction. This selectivity in barrier function was attributed to the superior competitive ability of species from a more diverse ecosystem. The newer data, referred to above, still indicate that it is much easier for animal species to migrate eastward across Wallace's Line than to go in the opposite direction. For plants, the data are not so unequivocal but, on the whole, the eastward movement appears to have been the greater.

Chapter 5

NORTHERN CONTINENTS SUMMARY

> Who can explain why one species ranges widely and is very numerous, and why another allied species has a narrow range and is rare? Yet these relations are of the highest importance, for they determine the present welfare, and, as I believe, the future success and modification of every inhabitant of this world.
>
> Charles Darwin, *On the Origin of Species,* 1859

It may be seen that, throughout their Mesozoic and Cenozoic history, the northern continents have either been joined to one another or have remained in close proximity. The most persistent marine barrier was the Turgai Sea which lasted from the mid-Jurassic to the Oligocene, about 130 Ma. The mid-Cretaceous formation of the Mid-Continental Seaway in the western part of Euramerica produced three northern land masses, the Beringia connection reduced the three to two, and the desiccation of the Mid-Continental Seaway reduced the two to one. The single northern continent that stood at the beginning of the Cenozoic lasted until the early Eocene when Europe and North America finally separated. The disappearance of the Turgai Sea in the Oligocene again produced a single land mass interrupted only by the North Atlantic sea barrier. The latter configuration has endured for most of the Cenozoic, altered by relatively brief inundations of the Bering Land Bridge.

The history of Beringia, and its function as a migration corridor, has received considerable speculation over the years. At first, the story seemed to be rather simple. It identified the movement of an Eocene Arcto-Tertiary flora out of Asia into North America along with an associated fauna. This eastward migration was supposedly followed by a gradual movement southward as the climate got colder. However, as more paleontological research was accomplished and as the phylogenetic relationships of individual plant and animal groups were worked out, a much more complicated picture emerged. The idea of relatively intact floras and faunas undergoing intercontinental migrations is no longer realistic. Instead, we can see that migratory success has differed greatly among the major groups (families and genera) and among the species that belong to those groups.

It seems clear that, as climatic conditions in Beringia changed, the vegetation changed and so did the associated animal life. The changing habitat produced migratory opportunity for some species and migratory inhibition for others. But, even during times when the land bridge was inhabited by a given community, it seems that species were actively moving back and forth. For example, when the area was covered by a temperate forest, several species of maples (*Acer*) moved from one continent to the other and back again. Other plant genera, various animal groups, and even the freshwater fishes showed the same tendency. This serves to remind us that it is the representatives of individual species that do the migrating and that they often do not move in concert with a whole ecosystem. Once the emigrant species reaches a new area, it may evolve to a new level of adaptation and then return to its homeland.

While Beringia served as a relatively easy migratory route, the Caribbean connection, in regard to the Antilles and to Central and South America, has been very difficult. The latter, throughout most of its history, served as a sweepstakes route where fortuitous or waif dispersal was the dominant mode. Central and South America could not be reached by any substantial numbers of the North American biota and vice versa until the isthmian connection was completed in the late Pliocene. The preponderance of geophysical opinion about the plate tectonics of the Caribbean region, which favors the concept of a Proto-Antilles that once occupied the area between North and South America, is not supported by the biological evidence.

The biotic relationships between North and South America, the phylogeny of the poeciliid fishes, and the evolution of the herpetofauna in Central America, seem to require the presence of an archipelago from the late Cretaceous through most of the Tertiary. If a Proto-Antilles had been in the position of the present Central America during the late Cretaceous and was then pushed eastward to the Bahamas Platform, it would have left an enormous oceanic gap in the Central American island chain. The creation of a new archipelago in the eastern Pacific and its subsequent movement to fill the gap would have taken considerable time, approximately the first half of the Tertiary. However, this is the time period for which we now have considerable evidence of waif dispersal between the continents, movements which must have been facilitated by the presence of islands.

Other evidence that refutes the Proto-Antilles concept lies in the characteristics and relationships of the Antillean biota and in the geology of the major islands. The vertebrate fauna is noticeably disharmonic with the island species representing those mainland groups that are most likely to disperse across saltwater. None of the animals are so old as to be considered Mesozoic relics. Investigations into the geology of the Greater Antilles indicates that their subaerial emergence has been comparatively recent, probably at least late Eocene. Thus the biological data suggest that the present Central America developed more or less in situ from an ancestral archipelago that was in place by the latest Cretaceous, and that the Antilles are oceanic islands that gradually accumulated their fauna and flora by means of fortuitous overseas dispersal.

In regard to the Indo-Australian connection, the very sharp faunal change that takes place across the few miles of saltwater that separate Borneo from Celebes and Bali from Lombok was, for many years, difficult to explain. The presence of a rich oriental fauna on Borneo and the islands to the west of Wallace's Line became easy to understand when it was realized that the island group was situated on the continental shelf which was exposed during the Pleistocene ice ages. But, this still did not explain why the nearby islands just east of Wallace's Line had such different animals. The mystery was not solved until it became known that Australian plate, with its peculiar fauna, had moved in only a few million years ago.

Of the vertebrate groups, it is the freshwater fishes that show the most striking change at Wallace's Line. In Borneo there are many families of primary freshwater fishes yet none of them have managed to reach Celebes which is only 65 miles away. The other oriental vertebrates have been more successful in making their way out the Malay Archipelago. Three lizard families, two turtle families, and six snake families have extended their ranges all the way to Australia. Many oriental bird families extend well along the archipelago and some have invaded continental Australia. Some, but fewer, bird groups have managed to migrate from Australia to the Oriental Region.

The placental mammals have extended their ranges across Wallace's Line to the east with considerable success. One family of rodents has been in Australia since the early Pliocene. But, none of the marsupials has succeeded in crossing the line in the opposite direction. Australia has been invaded by numerous oriental insects, spiders, and other invertebrate groups but, as with the vertebrates, there is little indication of reciprocal migration. The plant life of the Malay Archipelago also demonstrates a dual origin. Although most of the dispersal seems to have been eastward toward New Guinea, there are many examples of westward extensions into the Oriental Region.

The history of the North Atlantic connection has become clarified in recent years. It is now apparent that, from the late Paleocene to the early Eocene, there was a strong relationship between the mammalian faunas of North America and Europe. Similar relationships have been shown by salamanders, four lizard families, and three families of freshwater fishes. This information, plus that provided by the distribution of certain marine fossils, tells us that a North American – Europe terrestrial connection apparently persisted from the Mesozoic to the early Eocene. In contrast, recent geophysical maps have shown continuous sea passages separating the two continents since mid-Mesozoic time.

Although the general history of tectonic plate relationships in the northern hemisphere is reasonably well known and is helpful in assessing the overall relationships of the continents through time, the study of plate tectonics has not been able to focus with sufficient accuracy on the details of land mass connections and interruptions. Without biological data, it would be very difficult to establish the time when Europe and North America separated, the time when the Bering Land Bridge was first established, and to present a satisfactory history for the Greater Antilles and Central America. Moreover, it would have been impossible to detect the existence and to predict the biogeographic importance of the two great epicontinental seas.

Because of the past connections and the present proximity of the northern continents to one another, the terrestrial biotas of the temperate northern hemisphere are to a large extent homogeneous. During most of the Cenozoic, when many of the families and the great majority of our present genera and species evolved, there was a single northern land mass broken only by the ocean passage connecting the North Atlantic to the Arctic Ocean. The close relationships which may be seen among the non-tropical, northern hemisphere animals and plants have led many biogeographers to treat the entire area as a single Holarctic Biogeographic Region.

Part 2

THE SOUTHERN CONTINENTS

INTRODUCTION

There exists considerable confusion and disagreement about the history of the southern continents. As King (1980) remarked on the eve of the Fifth Gondwana Symposium which was held at Wellington, New Zealand, "For the solution of jigsaw puzzles, the first requirement is that one should have all the pieces. This is unhappily not so for Gondwanaland." Some of the missing pieces may be with us in the form of oceanic plateaus that represent detached continental fragments (Ben-Avraham, 1981), or it is perhaps conceivable that the earth has expanded some 20% in size since the Triassic (Owen, 1983).

It has been noted that the effects on the terrestrial biota of the great extinction that took place at the end of the Permian are not well documented. However, among the major tetrapod groups some important evolutionary changes took place that might be related to the continental assembly. All the major tetrapod groups which lived in the Paleozoic virtually died out during the Triassic. They comprised various sorts of ancient amphibians, a number of the most primitive reptiles, and many different groups of the mammal-like reptiles or synapsids (Charig, 1979). At the same time, all of the tetrapod groups that dominated the continents during the Mesozoic made their effective appearance during Triassic times.

From a biogeographic standpoint, it is important to note that an almost worldwide, interrelated vertebrate fauna existed by the end of the Triassic (Cox, 1974; Charig, 1979; Olson, 1979). In Africa, the Lower Triassic Middle and Upper Beaufort beds have yielded fossils in two zones that are named from characteristic reptiles (Colbert, 1980). These are the *Lystrosaurus* zone and the *Cynognathus* zone. These faunas are of considerable interest for they have subsequently been discovered in other places. A number of genera, known from the *Lystrosaurus* zone, have turned up in the Fremount Formation of Antarctica. Also, certain genera of the mammal-like reptiles of the *Cynognathus* zone have been found in the Puesto Viejo Formation of Argentina and in Australia (Thulborn, 1983). In central India, the Panchet beds have yielded both *Lystrosaurus* and some of the genera of the *Cynognathus* zone.

Some of the foregoing information led Colbert (1973) to conclude that firm evidence was at hand for the existence of a continuous Gondwanaland in the southern hemisphere. However, China then had to be added to the Gondwana confederation because of the discovery of a *Lystrosaurus* fauna there (Crawford, 1974) and, finally, the same fauna was found in the European part of Russia (Kalandadze, 1975). So, it now appears that the fauna of the two zones originally discovered in Africa offer better support for the presence of a north-south continental connection rather than a separate Gondwanaland. These facts appear to be reasonably consistent with the idea that a separate Gondwanaland (defined as an amalgamation of all the southern continents separated from Laurasia to the north) may not have existed in the Mesozoic. Additional information, pertinent to this question, will be presented as the history of the southern land masses is discussed.

In contrast to the historic proximity and close biogeographical relationships of the northern continents, the major southern land masses became separated at an early date and, in general, have remained isolated for extended periods of time. These periods of isolation had profound evolutionary effects on the various biotas. In each case, the composition of the contemporary and fossil flora and fauna can help us to understand the past. Since each southern land mass has its own unique characteristics, they are discussed one at a time.

Chapter 6

NEW ZEALAND

For over one hundred years, since Darwin and Hooker corresponded, biologists have been divided in their attempts to explain common biotic elements in Australasia, southern South America, and other southern lands. Some advocate land connections between these lands and an Antarctic center of origin, others rely on transoceanic dispersal, others again consider southern lands as refuges for organisms driven south and replaced in northern areas of origin.

C.A. Fleming, *Biogeography of the Pacific Basin: Antarctic Relationship,* 1961

The comprehensive geological history of New Zealand and its biota by Fleming (1975) is helpful for the determination of early biogeographic relationships. The earliest period for which fossil material is available is the Permian. At this time, according to most of the continental drift interpretations, New Zealand was supposed to be an integral part of the southern edge of Gondwanaland (itself forming the southern part of a larger Pangaea). This geographical position would place it in the coldest seas of that time since it is far removed from the tropics. Yet, Fleming was able to identify, among the Permian marine invertebrate fossils, two biogeographic elements that were to play a continuous role in the subsequent history of New Zealand. These were a cool, southern Austral element and a tropical or subtropical Tethyan element.

By the Triassic and Lower Jurassic, many brachiopod and molluscan genera were either confined to New Zealand or else shared with New Caledonia. This may, as Fleming (1975) pointed out, have been due to an isolation effect or the endemism may be only apparent since there is a scarcity of fossiliferous Triassic deposits in other parts of the southern hemisphere. In the Lower Jurassic, an isolation is indicated by the complete absence of belemnites and trigoniids which were then abundant in other parts of the world.

The New Zealand marine faunas of the Middle and Upper Jurassic demonstrate a strong affinity to the tropical Tethys. The Jurassic terrestrial flora is similar to that of the Triassic and shows a relationship to other parts of the southern world (Fleming, 1975). Included are ancestors of many primitive plants (lycopods, ferns, araucarians, podocarps) that persisted to Tertiary and Recent times.

One reptile, the tuatara (*Sphenodon*), and the single frog genus (*Leiopelma*) may have reached New Zealand in the Triassic or Jurassic. An abundance of rhyncocephalian fossils related to *Sphenodon* have been found in the Middle and late Triassic of Africa, Indian, and South America (Colbert, 1980). A fossil frog genus related to *Leiopelma* has been described from the late Jurassic of Patagonia (Estes and Reig, 1973). Craw (1985) felt that *Leiopelma* (Leiopelmidae) was related to the family Discoglossidae and thus showed direct transpacific relationships to the northern hemisphere rather than via Antarctica. But this differs from the survey of anuran evolution and biogeography by Savage (1973) who concluded that the leiopelmids originated in the southern hemisphere and the discoglossids in the northern hemisphere.

The history of the physical relationship of New Zealand to Australia is not at all clear. From a study of linear magnetic anomalies, Hayes and Ringis (1973) indicated that the Central Tasman Sea formed between 60 and 80 Ma and that the southern part of the sea was only about 50 Ma old. This indication that New Zealand and New Caledonia may have been attached to Australia as late as the early Tertiary may be contrasted with the assumption of Fleming (1975) that the New Zealand geosyncline began to break off from the rest of Gondwanaland at the end of the Jurassic.

The Cretaceous marine faunas of New Zealand were primarily of Tethyan or cosmopolitan affinities suggesting the presence of warm-temperate seas (Fleming, 1975). By the Cenomanian of the Middle Cretaceous, the bivalve molluscan fauna had become highly distinctive (Kauffman, 1979). The late Cretaceous dinosaur remains (Molnar, 1981) are a puzzle. They consist of caudal vertebrae (Russell, 1984). Could they have belonged to an aquatic creature? The first angiosperm pollen was reported from the Albian of the Lower Cretaceous in about the same horizon as in Australia and North America. As the angiosperms increased, many characteristic Cretaceous lower plants and gymnosperms became extinct. By the end of the Cretaceous, substantial generic elements of the living flora had appeared. Cracraft (1973) suggested that the history of the ratite birds extended well back into the Cretaceous and that their ancestors were flying birds living in Gondwanaland. If the kiwis and moas had a flying ancestor, it could have reached New Zealand over a water barrier.

The Cenozoic biotic history of New Zealand is consistent with a geographical position of extreme isolation. Its separation from other land masses is reflected in the continuous fossil record afforded by its shallow water marine fauna. The dual relationship between a cool Austral fauna and a warm Tethys fauna, which began in the Permian, continued throughout the Cenozoic. The austral forms became prominent at the end of the Cretaceous, a time of worldwide sea surface cooling, and then declined as warmer conditions began to prevail. The Tethys fauna gradually increased to a peak in the lower Miocene then decreased (Fleming, 1975). During the Tertiary, there was a steady influx of Australian species which increased considerably in the Pleistocene and Holocene. Despite the Lower Miocene climax of Tethys influence, a number of tropical genera were missing. This suggests that, since the Mesozoic, New Zealand has lain outside the tropical zone.

In recent years, various groups of the present New Zealand biota have been investigated to the extent that it is now possible to provide a reasonable estimate of its relationship to other areas. In the freshwater environment, the bivalve mollusc fauna is comprised of very primitive genera but the close relationship of the species to those of Australia indicates a recent colonization from the latter, probably by sea birds (McMichael, 1958, 1967). The ancient decapod family Parastacidae is represented in New Zealand by an endemic genus; the family also extends to Australia, New Guinea, Aru Islands, South America, and Madagascar (Bishop, 1967). *Paratypa curvirostris* is a freshwater shrimp of an old group that probably originated in the Tethys Sea during the Cretaceous (Carpenter, 1977); the genus occurs in an arc along the western Pacific from Japan to New Zealand; the most northern and most southern species appear to be the most primitive.

The mayfly (Ephemeroptera) fauna of New Zealand belongs almost entirely to two very old families, the Leptophlebiidae and the Siphlonuridae, both found primarily in the southern hemisphere. In the Leptophlebiidae, four genera are most closely related to New Caledonian genera, two are related to Chilean genera, and the remainder are closer to other New Zealand genera (Towns and Peters, 1980). In the Siphlonuridae, four of the five genera belong to subfamilies that also exist in southern South America and in Australia (McLellan, 1975). All New Zealand mayfly genera are endemic. There are also at least two monotypic subfamilies and one monotypic family. The stoneflies (Plecoptera) are represented by four families, all of them restricted to the southern hemisphere (McLellan, 1975, 1977); one subfamily, the Antarctoperlinae, is of particular interest because it is found only in New Zealand (10 species) and South America (14 species). All of the stonefly genera are probably endemic.

The dragonfly (Odonata), megalopteran, and mecopteran faunas are sparse and closely related to those of Australia (McLellan, 1975). The New Zealand caddisfly (Trichoptera) fauna consists of 14 families. Some are recent arrivals from southeast Asia, three are restricted to New Zealand and Australia, and one is endemic. The family Rhynchopsychidae is found only in New Zealand and Chile (one genus in each place); other trichopterid families indicate close relationships to both South America and New Caledonia (Cowley, 1978). The chironomid midges (Chironomidae) are not well known; two of the subfamilies appear to be very primitive and demonstrate closer relationships to South America than to Australia (Brundin, 1966). The midges of the family Blephariceridae have an interesting distribution in that there are relatively primitive genera in New Zealand and South America but more advanced genera in Australia (Zwick, 1977).

Only one genus of the dipteran family Simuliidae occurs in New Zealand; species in this genus are also found in Australia and one is in South America (McLellan, 1975). The two described species of New Zealand Thaumaleidae belong to a genus also found in Australia and southern South America. The freshwater fish fauna of New Zealand consists of 8 families, 10 genera, and about 30 species (McDowall, 1978). Most of the species, but only two of the genera, are endemic. This suggests that the fauna is not of great age. None of the families may be considered primary or secondary freshwater groups in the sense of Myers (1938). In general, they are saltwater tolerant and all apparently include some species that spend at least part of their life cycle in saltwater. The suborder Galaxoidei, which contains four New Zealand families, is a very primitive, southern hemisphere, bony fish group that may have become more widespread as the continents moved farther apart, but could just as easily have become dispersed by oceanic currents. The same dispersal scenario could be true for the other native freshwater fishes.

In the terrestrial environment, many of the very old groups of land snails are confined to a Southern Relict Zone which includes New Zealand and other parts of Oceania (Solem, 1979). This old fauna is primarily western Pacific with very little relationship to South America or South Africa, yet it must have been on New Zealand since the Cretaceous or even earlier. The land snail fauna of New Zealand is comprised of 9 families, 6 of which are considered to be very primitive; one is endemic (Climo, 1975).

The New Zealand arachnid fauna is, in general, quite closely related to that of Australia. There are many common species and species pairs suggesting a regular interchange (Forster, 1975). In some of the more primitive taxa of both harvestmen and spiders, there are southern hemisphere relationships that are apparently quite ancient. The spider family Archaeidae has been recorded from New Zealand, Australia, South Africa, and Madagascar and the family Mecysmaucheniidae is found only in New Zealand and South America. Also, some genera have broad distributions in the south but they cannot be nearly as old as the families themselves.

The terrestrial insect fauna of New Zealand is large and varied. In most groups, 90 – 100% of the species are endemic. There are many endemic genera and there is an endemic family in the Lepidoptera and another in the Coleoptera. Although a number of families and subfamilies are confined to New Zealand and Australia, many more have a broader paleaustral distribution including New Zealand, Australia, and southern South America. Both the New Zealand – Australia endemic families and the paleaustral families comprise mainly primitive groups. Some paleaustral groups apparently once had a worldwide distribution but became extinct in the northern continents. The family Sciadoceridae has one living species in New Zealand and Australia and another at the southern tip of South America. However, a fossil sciadocerid was recently discovered in the Oligocene Baltic amber and two others in Cretaceous amber from Canada. It appears that some, perhaps most, paleaustral groups originated in the north, but there is also abundant evidence that some originated in the south where they have remained. A few paleaustral groups are found in New Zealand and Chile but are absent from Australia. In contrast to the old elements, many groups of modern, dispersable insects show strong relationships at the specific level within the southern hemisphere (Watt, 1975).

The general relationships of the New Zealand flora have been summarized by Godley (1975). The fern flora has a low degree of endemism with about 50% of the species being found in southeastern Australia; of the New Zealand genera, about 85% are represented in Australia. In the seed plants, many of the species and 75% of the genera are shared with Australia. The close species relationships are attributed to recent trans-Tasman dispersal. On the other hand, many of the important Australian genera of Myrtaceae and Rutaceae and the Australian families Mimosacaceae and Casuarinaceae do not occur in New Zealand. Relatively close relationships with Chile are shown by a sharing of 40% of the genera of ferns and 43% of the genera of seed plants.

Galloway (1979) has noted that the southern hemisphere is the center of distribution of several lichen genera which are thought to have evolved there and which are represented in the northern hemisphere by only one or two species. The number of lichen species common to New Zealand and South America is small but there is a large number of genera in common, so that similar ecological niches in both countries are often filled by related species. The austral element of the New Zealand lichen flora is characteristic of undisturbed habitats and is well developed in the epiphytic vegetation of beech and podocarp forests, and in subalpine scrub associations and subalpine grasslands.

As Schuster (1972) has shown, a number of primitive bryophyte genera found in New Zealand have discontinuous distributions in the southern hemisphere; several occur in South America as well as in various parts of the southwestern Pacific. One species, *Saccognidium australe,* is found only in New Zealand and Chile. In the old conifer family Araucariaceae, the genus *Araucaria,* currently with a disjunct distribution only in the southern hemisphere and absent in New Zealand, did occur in New Zealand from the Jurassic to the Oligocene and apparently was also widespread in the northern hemisphere (Florin, 1963). A related genus *Agathis* (kauri pines) is present in New Zealand as well as in the general Indo-Australian Archipelago area. The Podocarpaceae with 7 genera and about 150 species is the most important conifer family in the southern hemisphere. Several of the genera and their subsections are found in New Zealand and demonstrate past relationships to the Indo-Malayan area, Australia, Antarctica, and South America; two subsections also occur in Africa. In the cedar family Cupressaceae, related genera are found on New Zealand and New Caledonia (*Libocedrus*), New Guinea (*Papuacedrus*), and Chile (*Astrocedrus* and *Pilgerodendron*).

In his review of the orders and families of primitive living angiosperms, Smith (1972) considered the Winterales, with the single family Winteraceae, to be the most primitive order. The distribution of the Winteraceae is primarily southern with six of the seven genera having relict distributions; one (*Pseudowinteria*) is restricted to New Zealand and no less than four (two endemic) are found in New Caledonia. The one beech (Fagaceae) genus of the southern hemisphere is *Nothofagus*; it is often considered to be a key genus in plant geography (van Steenis, 1972). By late Cretaceous times, it had apparently spread over an area including southern South America, Antarctica, Australia, Tasmania, and New Zealand; it also reached New Caledonia and New Guinea but perhaps not until somewhat later (Humphries, 1981).

The New Zealand terrestrial vertebrate fauna is extremely limited. As already mentioned, the archaic vertebrates consisting of the reptile *Sphenodon* and the frog genus *Leiopelma* probably date back to Triassic/Jurassic times. The late Cretaceous dinosaur remains consist only of some tail vertebrae. The two orders of ratite birds (moas and kiwis) are possibly Cretaceous in origin. There are three endemic genera of the lizard family Gekkonidae. The family is considered to have evolved sometime in the Mesozoic, probably during the Upper Jurassic/Lower Cretaceous (Kluge, 1967). The New Zealand genera belong to the primitive subfamily Diplodactylinae which is restricted to Australia, New Zealand, New Caledonia, and the Loyalty Islands. There are also two genera belonging to the lizard family Scincidae. However, the skink genera belong to the most specialized subfamily (Lygosominae) and the genera themselves are widely distributed in Australasia (Greer, 1970). It has been suggested that the geckos probably entered New Zealand in the Miocene and the skinks in the Pleistocene (Bull and Whitaker, 1975).

There are 65 species of native land and freshwater birds including the ratites. Of particular interest are three endemic families of passerines, the wattle-birds (Callaeatidae), the "thrushes" (Turnagridae), and the "wrens" (Xenicidae); they probably represent three original colonizations that took place during the early Tertiary (Bull and Whitaker, 1975). There are also two endemic genera of parrots and three endemic genera of waders. In general, the modern bird fauna is most closely related to that of Australia. During the last 120 years, 8 Australian birds have established themselves as breeding species in New Zealand. Eleven other species cannot be distinguished from Australian subspecies so are probably also recent colonists.

NEW ZEALAND SUMMARY

It is obvious that the New Zealand biota is in a disharmonic or unbalanced state and that it has existed in that condition for a very long time. Most biologists who have worked with New Zealand organisms agree that its isolation extends back to at least Cretaceous times. What about the earlier Mesozoic relationships? We have noted that the rhynococephalian relatives of *Sphenodon* were widespread in Africa, India, and South America. However, *Leiopelma* affords a better clue since a related genus has been described from the late Jurassic of South America and that continent may have served as an important center of evolutionary origin for the frogs of the world (Briggs, 1984b). The late Cretaceous dinosaur remains are fragmentary. If the ratite birds had a flying ancestor as Cracraft (1973) suggested, the tinamous (Tinamidae) of South America may be closest to the ancestral form.

Some of the old aquatic insect groups demonstrate a restricted, relict distribution that emphasizes the New Zealand – South American relationship. In the mayfly family Leptophlebiidae, there are generic relationships only with New Caledonia and Chile, the stonefly subfamily Antarctoperlinae is found only in New Zealand and Chile, and the caddisfly family Rhynchopsychidae is also restricted to New Zealand and Chile. Two New Zealand subfamilies of chironomid midges demonstrate a closest relationship to South America. A few terrestrial insect families are found only in New Zealand and South America.

The cedar family Cupressaceae has related genera on New Zealand, New Caledonia, New Guinea and Chile. By the late Cretaceous, the southern beech *Nothofagus* apparently occupied an area including New Zealand, southern South America, Antarctica, Australia, and Tasmania. A strong affinity between the older elements of the New Zealand flora and fauna and those of southern South America is evident. In contrast, the relationship of the ancient biota to that of the rest of the southern hemisphere (Australia, Africa, Madagascar, India) is relatively weak.

Chapter 7

AUSTRALIA

Nothing in biology makes sense except in the light of evolution.

Theodosius Dobzhansky, *The American Biology Teacher,* 1972

In considering the historical location of continental Australia, the nature and age of the sea floor between Australia and Antarctica and the relationships of its old, shallow water marine fauna are of paramount importance. Magnetic anomaly data appear to indicate that a rift zone began to open between Australia and Antarctica in the late Jurassic and that deep sea conditions extended along most of the rift by 80 Ma (Cande and Mutter, 1982). Zinsmeister (1979) found an overall similarity of the marine molluscan faunas from the Cretaceous of southern South America to southeastern Australia indicating that the region comprised a single, broad faunal province; it disappeared by the late Eocene. Hotchkiss (1982) found that the distribution of Paleogene echinoids suggested a connection between Antarctica and South America but implied a major barrier between Antarctica and Australia.

By the late Permian, continental assembly had taken place to the extent that new reptile lines, particularly the therapsids, had become widespread in the north and had also invaded South America, Africa, Madagascar, and India. Triassic fossils of this fauna have been found in Antarctica and Australia (Cox, 1974). In the Triassic, several terrestrial reptiles inhabited Australia (Molnar, 1984). These included a proterosuchian, a member of the oldest group of archosaurs that eventually gave rise to the saurischian and ornithischian dinosaurs, and reptiles of the *Lystrosaurus* assemblage. By the Jurassic, there were plesiosaurs, sauropods, and theropods. Also a labrinthodont amphibian persisted into the Lower Jurassic. In the Cretaceous, there were a large variety of reptiles including ornithischians, sauropods, theropods (with a large *Allosaurus*), and pterosaurs.

In his summary of Australian reptilian relationships during the Mesozoic, Molnar (1984) noted that there was evidence for free passage of land forms into Australia during the Triassic. But the Jurassic terrestrial faunas are composed entirely of peculiar and late-surviving forms that had already become extinct in other parts of the world. In the Cretaceous, it is the aquatic forms that resemble those from overseas while the land forms, by and large, are either endemic or late survivors. These are indications that unobstructed terrestrial passage into Australia became disrupted at the end of the Triassic.

The modern herpetofaunal relationships between Australia and South America have been reviewed by Tyler (1979), Tyler et al. (1981), and by Cogger and Heatwole (1981). The diverse frog family Leptodactylidae is found in South America (three subfamilies), Australia (two subfamilies), and southern Africa (one subfamily). The South American affinities of the Australian group have been clearly demonstrated. In a similar manner, the members of the treefrog family Hylidae in Australia and South America are also related, those of each area belonging to a different subfamily. The freshwater turtle family Chelidae occurs in both regions; there are seven genera in South America and four in Australia. The latter group has undergone most of its evolution in South America and probably reached Australia in the late Cretaceous or early Tertiary.

Although it has been suggested that the ancient lizard family Gekkonidae radiated from a Neotropical evolutionary center (Tyler, 1979; Briggs, 1984a), Richard Estes (by letter) has pointed out that there were primitive gekkos in the Jurassic and Cretaceous of northern Europe and Asia. It is Estes's opinion that the family has a center of origin in southeast Asia. The large Australian gekko fauna belongs to a relatively primitive subfamily (the Diplodactylinae) that has no evident South American relationships. The discovery of primitive xiphodont crocodiles from the Pleistocene of Australia (Hecht and Archer, 1977), evidently related to a genus from the Eocene of Patagonia, may provide more evidence for an Australian – South American relationship. Although the foregoing examples are important, it should be borne in mind that they represent only a small fraction of the rich South American herpetofauna that evidently existed in the Cretaceous and early Tertiary.

Following the early infusions of reptiles and amphibians from South America, Australia in the mid-Tertiary began to pick up herpetofaunal contributions from the Oriental Region. These include the frog families Ranidae and Microhylidae; the lizard families Varanidae, Scincidae, and Agamidae; the turtle families Carettochelyidae and Trionychidae; and the snake families Elapidae, Colubridae, Acrochordidae; Uropeltidae, Typhlopidae, and Boidae (Tyler, 1979). Some of these families are new arrivals having so far penetrated only to the northernmost part of Australia while others, such as the Elapidae (25 genera and 63 species in Australia), are broadly distributed and have certainly been there for a long time.

In the avian fauna, the cassowaries, emus, and extinct dromorthinids (Rich, 1980) of Australia constitute an ancient group called the ratite birds. There is a distinct but somewhat distant (ordinal) relationship to similar large, flightless, recent and extinct ratites of New Zealand, Africa, Madagascar, and South America. Eocene ratite fossils have been recorded from South America, North Africa, and Switzerland. Cracraft (1973, 1980), on the basis of morphological and biochemical studies, decided that all the ratites had a common ancestry and that the South American tinamous were closest to the ancestral type. Cracraft felt that the ratites developed on Gondwanaland and attained their present distribution as the result of continental drift. However, the fact that they were present in Europe during the Eocene might mean that their early evolution took place in the north and that they retreated to the southern hemisphere because of competitive pressure from the modern birds or mammals.

The bird order Galliformes is also relatively primitive. Cracraft (1973) considered the Australian mound birds (Megapodidae) and the Neotropical curassows (Cracidae) to be the most primitive galliform taxa. He proposed a distributional history of the group that envisioned primitive galliforms developing in Gondwanaland then migrating from South America to North America and, from there, to Asia and Europe. However, the most successful galliform family, the Phasianidae, reaches its greatest diversity in the Oriental Region and this, plus the arrangement of the other taxa in the order, indicates that the center of origin may be located in the latter area (Darlington, 1957; Mayr, 1946).

The parrots (Psittacidae) reach their greatest generic diversity in the vicinity of Australia (Keast, 1981). Although Cracraft (1973) assumed that their early history took place in Gondwanaland, the existence of several peculiar subfamilies in the general Australian region (including New Zealand) indicates a long evolutionary history in that part of the world. In a similar manner, the distribution of the pigeons (Columbidae and allied groups) also indicates an extended history in the Australian area. Considering our present knowledge of the two families, it seems reasonable to suggest that Australia has been the center of evolutionary radiation for both parrots and pigeons.

Rich (1975, 1979) has published two accounts of the past and present relationships of Australian birds. She considered about 43% of the living, non-passerine fauna to be of uncertain geographic origin and that the remaining groups probably arrived via the Indo-Australian Archipelago. She concluded that the avian fossil record can offer neither support for, nor criticism of, current Gondwanaland reconstructions. More recently, Sibley and Ahlquist (1986) have been able to shed additional light on bird phylogeny by using a DNA-DNA hybridization method. It was discovered that, among the songbirds which belong to the suborder Passeres, there are two major groups, the Passerida and the Corvida, that diverged from a common ancestor about 55 – 60 Ma. The Passerida evolved in Africa, Eurasia, and North America while the Corvida evolved in Australia. Beginning about 60 to 30 Ma, when Australia was most isolated, the Corvida evolved into many specialized forms including warblers, flycatchers, creepers, thrushes, etc. The resemblance to northern families was so great that the Australian species were orginally placed in those groups.

Perhaps the most startling result of the research by Sibley and Ahlquist (1986) was the discovery that the Australian endemics were most closely related to one another and that their resemblance to northern families was the result of convergent evolution. Furthermore, the Corvida of Australia produced the ancestors of some groups that were able to emigrate to Asia as Australia moved northward. This produced an initial radiation in southeast Asia which then extended to most other parts of the world (Fig. 9). So it seems clear that our modern species of the family Corvidae (crows, ravens, jays, magpies, etc.) can trace their ancestry to Australia. The crows and ravens of the genus *Corvus* probably originated in Eurasia, became widespread, and eventually returned to colonize Australia.

The mammalian relationships provide some interesting parallels to those shown by the herpetofauna. Although the earliest marsupials might have been more widespread, it is probably significant that Cretaceous marsupial fossils are known only from North and South America. In the Eocene, marsupials reached Europe and by the early Oligocene had penetrated to Asia and North Africa. But by the Miocene they were extinct in the northern hemisphere (p. 87).

It was in South America that the marsupials enjoyed their longest tenure – from the beginning of the Cenozoic until recent time. And, according to the modern concensus, it was from South America that some marsupials invaded Australia in the early Tertiary by way of Antarctica (Colbert, 1980; Archer, 1984). In a serological study, Kirsch (1977) found that the Australian marsupial taxa were closely related to one another indicating that they had all evolved from a few immigrant forms. In contrast, the greater differences among the South American forms suggested a longer evolution in that area. He felt that all of the events of marsupial evolution in Australia had taken place within the last 50 – 60 million years. The recent discovery of the polydolopid marsupial from the late Eocene of Antarctica, related to South American taxa that were abundant about 50 Ma (Woodburne and Zinsmeister, 1984), is consistent with the prediction by Kirsch.

Before being completely isolated from North America, South America possessed a primitive placental fauna which rapidly developed into a great variety of forms (liptoterns, notoungulates, condylarths, xenungulates, edentates, etc.), yet the Marsupialia was the only one of several diverse mammalian orders with a long history in South America to make it to Australia (Simpson, 1980). Although it is commonly thought that marsupials predated the placental mammals and, therefore, may have gotten to South America first, the current concept is that both groups arose from their eupantothere ancestral stock at the same time (Colbert, 1980). Patterson (1981b) has noted that both existed in South America for an equal time.

The strange monotremes, the group to which the Australian platypus (*Or-*

Fig. 9. The early evolution of the bird tribe Corvini (crows, ravens, magpies, and their relatives) was confined to Australia – New Guinea. Then, as that continent approached the Malay Archipelago in Oligocene/Miocene times, ancestral forms reached southeast Asia where a secondary burst of evolutionary radiation took place. Modern crows and ravens have reinvaded Australia from Asia. Redrawn after Sibley and Ahlquist (1986).

nithorhynchus) and the spiny anteater (*Echidna*) belong, are intermediate between the mammal-like reptiles and modern mammals. Colbert (1980) noted that the shoulder girdle of a docodont (one of five orders of early mammals), recently recovered from a Triassic deposit in New South Wales, is closely comparable to that of the monotremes. An Australian monotreme from the early Cretaceous has been described (Archer et al., 1985). These may be indications that monotremes have been in the Australian region throughout their history.

Placental rodents of the family Muridae (about 13 genera) are abundant in Australia. It seems clear that they reached there by island hopping down the East Indian chain. While Simpson (1965) suggested that the first arrivals might have entered Australia in the Miocene, more recent research on murid relationships (Hand, 1984) seems to indicate a later date, the early Pliocene. The oldest rodent fossils are 4 – 5 million years in age. Invasions by the genus *Rattus* evidently took place in the Pleistocene. Throughout, the predominant movement has evidently been from the smaller East Indian islands to the great island of New Guinea and then from New Guinea to the continent of Australia.

The true freshwater mussels of the superfamily Unionoidea are a very old group that should be helpful in determining continental relationships. The Australian species belong to the family Hyriidae; within this family, the Australian genera are placed in two subfamilies and certain South American genera are placed in another. The resemblance in anatomy among the three subfamilies and the similarity of their larval stages is considered to be indicative of an ancient common ancestry (McMichael, 1967; Walker, 1981). The snail family Hydrobiidae has endemic genera in Australia that also may be part of its ancient fauna. However, the several other families of freshwater molluscs are not highly endemic and are generally related to widespread Asian and Pacific island groups.

The freshwater crayfish family Parastacidae is a product of the nephropsidian radiation that occurred in the early Mesozoic (Glaessner, 1960). The present distribution of the family (Bishop, 1967) includes southern South America (*Parastacus*), New Zealand (*Paranephrops*), Australia (10 genera), New Guinea (*Cherax*), Aru Islands (*Cherax*), and Madagascar (*Astracoides*). This family may have originated in Australia and could have contributed a stock to South America and Madagascar in the early Tertiary (Riek, 1972). The one species of the family Hymenosomatidae has apparently invaded from the sea and the remaining three freshwater decapod families are circumtropical and probably got to Australia via the East Indies and New Guinea. In the bathynellid Crustacea, Schminke (1974) found a relationship linking Australia, New Zealand and South America.

Among the freshwater insects, the archaic mayfly (Ephemeroptera) family Leptophlebiidae has a predominately southern, cool temperate distribution. Recent systematic work on the South American genera indicated that they are most closely related to genera from Australia, followed by genera from New Zealand, Africa, and Madagascar, in that order (Pescador and Peters, 1980). Edmunds (1972) felt that mayfly distribution offered overwhelming evidence that the closest southern continent relationships were between Australia and southern South America (Fig. 10). In the stoneflies (Plecoptera), five families are restricted to the southern hemisphere; of these, three are shared only among Australia, New Zealand, and

South America; one extends also to South Africa and Madagascar; and one is endemic to South America (Zwick, 1981a).

In the caddisflies (Trichoptera), Ross (1967) noted that the Australian fauna consisted of an older group related to forms in New Zealand and South America and a group of modern genera from southeast Asia. It appears that there may have been two trichopteran dispersals to the three areas, an early one revealed by the presence of separate genera in each place and a later one identified by the presence of separate species in each place that belong to common genera. Neboiss (1977) called attention to several families with trans-Antarctic patterns.

Recent research on the Blephariceridae (Zwick, 1981b) revealed that the subfami-

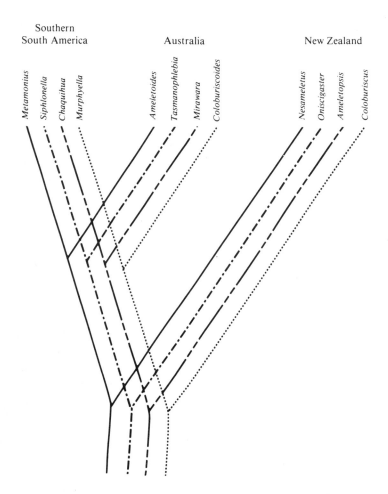

Fig. 10. Distributional relationships among four phyletic lines in southern hemisphere mayflies (Ephemeroptera). This cladogram indicates that the South American taxa are more closely related to those of Australia instead of New Zealand. From Brown and Gibson (1983) after Edmunds (1981).

ly Edwardsininae was represented by two genera in South America, one genus in Australia, and one in Madagascar. The dragonfly family Petaluridae, an ancient group that was dominant during the Jurassic, now contains only nine living species in five genera, *Petalura* (4 species) in Australia, *Uropetala* (1) in New Zealand, *Phenes* (1) in Chile, *Tachopteryx* (1) in eastern North America, and *Tanypteryx* (2) in western North America and Japan (Watson, 1981). The origins of the Australian waterbug (Hemiptera) families are probably oriental with the possible exception of a primitive genus of Corixidae that is shared with New Zealand (Lansbury, 1981).

An overview of the geographical relationships of the Australian aquatic insect fauna was provided by Williams (1981) who noted that the bulk of the fauna consisted of a southern element of primitive forms with austral (Gondwanaland) affinities and a younger, northern element of oriental affinities. The latter element was observed to have speciated greatly but not to have radiated much above the species level.

At one time, it was thought that the freshwater oligochaetes would offer good material for biogeographical studies but they have proved to be surprisingly adept at getting from one place to another (Timm, 1980). Fifteen species are found on all six continents and another 17 species exist on five continents. The one family almost confined to the southern hemisphere occurs on six isolated sub-Antarctic islands as well as on the southern continents and Sri-Lanka.

The native freshwater fish fauna of Australia has probably been entirely derived from the surrounding marine environment. There are no primary or secondary freshwater fishes in the sense of Myers (1938). The fact that Australia has a lungfish (*Neoceratodus*) and an osteoglossid (*Scleropages*) has often been cited as evidence of Gondwanaland connections for that continent. However, at least one extinct osteoglossid was marine (Patterson, 1975) indicating that the group may have had a marine origin. Fossil lungfishes, belonging to genera related to the modern *Neoceratodus,* have been found in Triassic marine beds (Patterson, 1975) and many Paleozoic lungfishes were marine (Thomson, 1969). If Australia had been close to another continent since the Upper Jurassic, one would expect to at least find some secondary freshwater fishes, which are salt tolerant to the extent that they can cross minor oceanic barriers, but there are none. The one possible candidate for this kind of relationship is a fossil species of the presumed freshwater family Archaeomaenidae that was found in the Lower Jurassic of Australia and Antarctica (Schaeffer, 1972).

In his work on the distribution and ecology of the Stylommatophora, the dominant group of land snails, Peake (1978) emphasized that very few families exhibit southern disjunct distributions with representatives shared between South America, Africa or Australia and none have representatives on all three continents. Five Australian families are considered to be members of the southern fauna; of these, two are shared with South America (one extending northward to North America), one with Africa, and two are endemic. The remainder of the Australian land snail fauna consists of seven families that are very widespread, two that have a Polynesian distribution, and one that is primarily found in the northern hemisphere.

In regard to the terrestrial oligochaete worms, Jamieson (1981) felt that the Australian fauna was largely derived from successive Gondwanan invasions mainly in the family Megascolecidae. A much smaller endemic component was the result of post-Miocene invasions from the Oriental Region. The Megascolecidae, as most other earthworm families, is very widely distributed. Although all four of the sub-families have a predominant southern hemisphere distribution, it would be difficult to defend a Gondwanaland origin for all occur to some extent in the northern hemisphere. The very primitive genus *Acanthodrilus* is restricted to New Caledonia but it must also be noted that some acanthodriles are euryhaline and could have been transported by sea.

Considerable information on the distribution of Australian terrestrial insects is available. Evans (1959) called attention to the importance of the ancient family Peloridiidae (Homoptera), a group of bugs that live on wet moss and cannot fly. Species belonging to several genera have been recorded from Chile, Patagonia, Tierra del Fuego, New Zealand, Tasmania, and Lord Howe Island. Paramonov (1959) particularly noted the close relationship in the old dipteran fauna between Australia and southern South America and listed 10 families that illustrate such ties. Schlinger (1974) found some interesting relationships among insects of the southern beech *Nothofagus* in which aphids of the genus *Sensoriaphis* are found in Australia, New Zealand, New Guinea, and Chile; a parasite of the aphid was found in Australia only. Another aphid genus, *Neuquenaphis,* occurs only in southern South America and it has both primary and secondary parasites.

The butterflies (Lepidoptera) are an old insect group in Australia. Tindale (1981) has reported lepidopteran wings from the Upper Triassic. Many of the Australian groups are considered to be archaic relicts. Several such groups demonstrate Australia – South America relationships. The most commonly recurrent generic pattern present among the older elements of the major beetle (Coleoptera) families in Australia is one depicting south temperate relationships, particularly between Australia and South America (Howden, 1981). Such genera are often endemic to Australia with sister genera occurring elsewhere. Typically, there are no close relatives in the Oriental or Palearctic Regions. The time of origin of most of these groups is probably Jurassic to early Cretaceous.

A beetle tribe (Migodopini) is confined to the southern hemisphere with four genera in Australia – Tasmania, four in New Zealand, and seven in South America (Thiele, 1977). However, this does not necessarily mean a southern ancestry, for the tribe is closely related to the more primitive Elephrini which is confined to the northern hemisphere.

In the termites, a relatively well known group because of its economic importance, the Mastotermitidae is the most primitive family; an Australian species is the only living member but fossil species, belonging to the same genus, have been described from the Tertiary of Europe and North America and a related fossil genus is known from Brazil (Krishna, 1970). In general, the older genera of the more archaic families tend to have relict distributions in the temperate parts of the northern and southern hemispheres. The Hodotermitidae is a primitive family with two genera; one has species in Australia, New Zealand, and South Africa while the other has species in Australia, South Africa, and South America (Emerson, 1955; Calaby and Gay, 1959).

In their monograph on the flat bugs (Hemiptera, Aradidae), Usinger and Matsuda (1959) noted that the three most primitive subfamilies had a southern temperate distribution; the Isoderminae inhabit the *Nothofagus* forest of South America, Australia, and New Zealand; the Prosympiestinae are found in Australia and New Zealand; and the Chinameyersiinae are confined to New Zealand. The robber flies (Asilidae) comprise a large family with over 400 genera and subgenera (Hull, 1962); it probably originated sometime in the Cretaceous and has undergone most of its evolutionary development in the Neotropical Region. One relatively primitive group (Chrysopogonini) is confined to the Australian Region and another (Phellini) is shared between Australia and Chile. Michener (1979), in discussing the geographical distribution of bees, noted that the primitive tribe Paracolletini occurs in Australia, South Africa, and temperate South America.

The fleas (Siphonaptera) are an old group that probably arose in the Jurassic (Traub, 1972). There are ancient subfamilies or tribes on the southern continents and three of the Australian families exhibit solid phylogenetic ties to South America. The modern Australian flea fauna immigrated from the north. In his studies on the Australian Aeolothripidae (Thysanoptera), Mound (1972) suggested that the group was originally distributed through a southern continent. In his summary on the composition and distribution of the Australian insect fauna, Mackerras (1970) called attention to a southern element that formed a significant primitive component of most orders of insects from mayflies to beetles. The fossil evidence suggests that it is composed of groups that had probably evolved into recognizable entities by the middle of the Mesozoic. Most of the included groups are shared between Australia and South America, fewer with South Africa or New Zealand.

The spider fauna of Australia consists of 46 families, most of which are widespread in the world (Main, 1981). However, a few of the families are primarily southern hemisphere and one, the primitive Actinopodidae, is found only in Australia and the Neotropical Region. In the family Archaeidae, the recent distribution is entirely southern hemisphere but fossils have been found in the Palearctic. Main suggested that all the widespread families probably attained cosmopolitan distribution during Pangaean times and that some have subsequently become more restricted due to changes in configuration of the continents and climatic zones.

According to Koch (1981), the composition of the scorpion fauna of Australia reflects the early indirect connection of this land mass to South America and the subsequent long isolation of Australia until its impingement upon Asia. The various genera of the scorpion family Bothriuridae are known only from Australia and South America but other scorpions in Australia are in the families Buthridae or Scorpionidae and belong to southeast Asian genera or to taxa that have clearly evolved from such genera.

Many groups of the more primitive terrestrial plants have been studied. Schuster (1969, 1981) called attention to a high concentration of primitive, relict genera of liverworts (Hepatophyta) in the Antipodes area (New Zealand, Tasmania, southeastern Australia, New Caledonia), and a significant, but lesser, number in southern South America. The distribution of the conifer and taxad genera in time and space have been analyzed by Florin (1963). One of the oldest of the living families is the Araucariaceae. The living species of *Araucaria* have an interesting disjunct distribution in the southern hemisphere (South America, Australia, New Guinea, Norfolk

Island, and New Caledonia). But, fossil material indicates a much wider distribution in the Mesozoic and early Tertiary. In the southern hemisphere, it once occurred in Antarctica, South Africa, New Zealand, Tasmania, and Kerguelen Island. To the north, it apparently ranged widely in both the Nearctic and Palearctic regions. The one other living araucarian genus (*Agathis*) ranges widely in the Western Pacific from the Philippines to New Zealand.

The Podocarpaceae with seven genera and about 150 species is the most diverse conifer family in the southern hemisphere (Florin, 1963). The various genera indicate a relationship among all the large land masses and, by means of fossil material, Antarctica and India. There is also fossil material from Kerguelen Island and living species on New Caledonia and Fiji. The genus *Podocarpus* is the largest. It is found not only in the foregoing areas but in the New World it extends northward to Mexico and the West Indies. In the Old World, it is found north to the Ryukyu Islands and Japan. The genus is subdivided into eight sections, seven of which occur in the East Asian – Australian – New Zealand area.

The family Taxodiaceae is predominately northern but one genus, *Athrotaxis*, lives in Tasmania and is fossil in Australia and New Zealand (Florin, 1963). The family Cupressaceae contains 22 genera, half of which are found in the northern hemisphere and half in the southern. Of the southern genera, seven are confined to the Australian – New Zealand – New Caledonian region, three to southern South America, and one to southern Africa. In his monograph, Florin presented a table showing the location of relict conifer and taxad genera. This shows a concentration of such genera in east and southeast Asia (13) and in the Australian – New Zealand area plus associated islands (13). In contrast, relatively few such genera are found in the rest of the world; southern South America (5), western North America (2), Europe (2), and Madagascar (1).

In regard to the angiosperms, Schuster (1972, 1976) has summarized much of the information on the distribution of the most primitive dicot families: of a total of the 19 most primitive families, 10 occur in the southeast Asian area (5 are endemic there) and 8 are concentrated in the Australian – New Zealand area (where 7 are endemic). The one southern family that is non-endemic to the Australian – New Zealand area is the Winteraceae. Smith (1972) considered the Winteraceae to be the most primitive angiosperm family. Of the seven genera, six are restricted primarily to the Australian – New Zealand area including New Caledonia, Lord Howe Island, and the Solomon Islands. One of the six is also found in Madagascar. So, from southeast Asia to the general Australian – New Zealand region, there is a remarkable concentration of primitive conifer and angiosperm genera and families.

It has been noted (p. 65) that, by late Cretaceous times, *Nothofagus* had apparently spread over an area including southern South America, Antarctica, Australia, Tasmania, and New Zealand (Humphries, 1981). It apparently never reached Africa or India. Schuster (1976) felt that the genus evolved in North America, migrated southward to Tierra del Fuego and from there via Gondwanaland connections to its present locations. Van Steenis (1972), on the other hand, emphasized the fact that all Fagaceae genera border on or occur in an area extending roughly from Sino – Himalaya – Hainan – Formosa to New Guinea and that the various genera must have evolved in that area and, from there, gradually dispersed to other parts of the world. Humphries (1981) concluded that *Nothofagus* was a wholly southern group whose precise center of origin was impossible to locate.

The distribution of *Nothofagus* undoubtedly offers useful biogeographic data, yet is not easy to interpret. Other information bearing on the matter, is that the South American *N. allessandrii* is considered the most primitive living species in the genus (van Steenis, 1972). This may be significant for, in many cases, the most primitive species is located farthest from its place of origin. Schuster (1972), in defending an American origin for the genus, emphasized that *Nothofagus* in South America is symbiotic with a hemiparasitic, unigeneric family of angiosperms, the Myzodendraceae, and that this relationship does not occur elsewhere in the range of *Nothofagus*. As noted earlier (p. 74), the presence of an endemic aphid genus, associated with *Nothofagus*, with both primary and secondary parasites (Schlinger, 1974), is also indicative of an extended age for *N. allessandrii* but does not necessarily mean that it was always confined to South America.

In an article on pre-Tertiary phytogeography and continental drift, Smiley (1979) called attention to some valuable facts. A late Paleozoic flora from New Guinea contains a mixture of taxa from the Cathaysian province of southeastern Eurasia and from the *Glossopteris* flora of Australia. Such a mixture would be expected if the two regions were in their present proximity but not if Australia were located far to the south. Smiley also referred to an early Cretaceous floral sequence near Melbourne in southern Australia that is similar to other early Cretaceous assemblages at present latitudes of 40 – 50°N, rather than to floras of the same age from 60° to 70°N. He felt that such evidence suggested an early Cretaceous position for southern Australia near its present latitude of 40°S.

In her summary on the phytogeographic relationships of the Australian Cretaceous flora, Dettmann (1981) concluded that significant alliances existed between Australia and other regions in southern Gondwanaland. The early Cretaceous microfloras from South America, India, New Zealand, and Australia share many common features. Although the initial radiation of the angiosperms evidently took place in the Tethyan Sea area in the early Cretaceous (Barremian), they did not reach Australia until the mid-Cretaceous (Albian) about 100 – 108 million years ago. Dettmann suggested that dispersal routes into Australia would probably have involved Antarctica, for which early angiosperm history is not yet documented.

In discussing the general relationship of the southern, cool temperate floras, Moore (1972) noted that between 50 and 60 genera, or distinctive parts of more widespread genera, occur principally in southern South America and Australasia, about half of them in New Zealand alone. Although some of these genera have significant developments in both the New and Old Worlds, most of the remainder appear to have migrated from east to west. The latter genera are those that have their major diversity in the Old World. The differentiation of such families as the Myrtaceae, Proteaceae, and Restionaceae apparently took place in the Cretaceous and their major evolutionary radiation evidently occurred in Australia. In the Proteaceae, all five of the subfamilies and most of the tribes are represented in Australia (Johnson and Briggs, 1975). Considering that the origin and early development of the angiosperms probably took place in southeastern Asia (Takhtajan, 1969; Smith, 1973), there would have had to be a migration route available leading to the

Australian – New Zealand area. Moore also noted that considerable recent or current intercontinental migration of plant species is taking place. Some 32 species, excluding cosmopolitan forms, occur in both the Australian and South American cool-temperate regions.

AUSTRALIA SUMMARY

There is in Australia a distinctive old fauna and flora descended from ancestral types that first arrived in the Mesozoic to early Tertiary. Such ancient elements can be found in group after group of the living biota; mammals; birds; reptiles; amphibians; freshwater animals such as mussels, crayfish, and insects; terrestrial invertebrates such as insects and snails; and plants such as liverworts, pinetrees, cedars, and many families of angiosperms. Following the acquisition of the old fauna and flora, there seems to have occurred a hiatus, a time when there were very few arrivals, dating from the early Tertiary to about mid-Miocene times. From the latter time onward, there was an influx of tropical organisms from the southeast Asian area via New Guinea. This new immigration, slow at first, appeared to pick up as time went on.

Important questions about the old fauna and flora are: (1) where did it come from, (2) how did it get to Australia? Fortunately, there are some unequivocal answers to the first question. Almost certainly, the marsupial mammals came from South America but the monotremes may have developed in situ. The leptodactylid and hylid frogs apparently came from South America as did the chelid turtles and probably the gekkonid lizards. The ancestors of the Australian ratite birds (cassowaries and emus) probably came from South America. Other imports from South America probably include the older elements of many common groups such as the aquatic mayflies, stoneflies, and caddisflies and the terrestrial carabid beetles, robber flies, and bees.

The other main source area for the old biota has evidently been the southeast Asian region. The general phylogenetic pattern in the primitive liverworts indicates that they probably spread from an evolutionary center in southeastern Asia. The same may be true for the Podocarpaceae and the Taxodiaceae. It seems apparent that the angiosperms also first evolved in the lowland tropics of the Oriental Region and gradually dispersed from there. As angiosperm evolution proceeded from the Cretaceous onward, the most primitive families retreated to the high altitudes of the mainland and also out the island chain of the Indo-Australian Archipelago. They must have reached the Australian – New Zealand area by the late Cretaceous. From there, some of them managed to spread to Antarctica and thence to South America. This is the route that must have been followed by the primitive dicot family Winteraceae and the southern beech *Nothofagus*.

Although there is very little evidence of an old relationship to Africa or India, the termites may have come from the former since Africa has been an important center for termite evolution. Finally, there are some groups for which Australia and its adjacent islands may have served as a center of evolutionary radiation, thus providing species to other continents. An interesting case among the birds is found in the family Corvidae to which the crows, ravens, magpies, jays and their relatives belong. The corvids probably began their development in Australia some 55 – 60 Ma. After Australia moved northward, some of the corvids invaded southeast Asia and underwent a further radiation. From there, they dispersed to most of the rest of the world. Other probable examples are the parrots, pigeons, the crayfish family Parastacidae, and the angiosperm families Proteaceae, Myrtaceae, and Restionaceae. As indicated in the text, there are many other old groups that demonstrate early intercontinental relationships but not enough is known about their history to give a reliable indication of their source areas.

The answer to the second question is more difficult. In looking at past continental relationships, the biogeographer should attempt to explain not only how some groups managed to make such journeys but why other groups did not. Both sides of the coin are important. For example, why did only a few marsupials manage to reach Australia from South America leaving other diverse, mammalian orders behind? Why did only a few of the amphibians and reptiles make the same trip? Why did the South American primary and secondary freshwater fish faunas completely fail to reach Australia?

The overall relationships of the old biota tells us that Australia must have been close enough to Antarctica, when the latter was larger (undepressed by an ice mantle) and when it supported a cool-temperate fauna and flora, to exchange organisms. How close? Close enough so that storm-driven rafts of vegetation with a few small animals aboard could occasionally (once every thousand years?) make the crossing with a chance to survive on the other side. But not so close that aquatic placental mammals could make it and not so close that secondary freshwater fishes, that can cross minor ocean barriers, could make it. Reptilian relationships demonstrated by Mesozoic fossils seem to indicate a free passage to and from Australia during the Triassic. But, except for aquatic forms, this communication was interrupted by the Jurassic and Cretaceous.

Most of the other old groups are, if one looks at worldwide patterns, more or less adept at dispersal over water. The ratite birds probably had a flying ancestor (similar to the South American tinamous). Freshwater molluscs and crustaceans or their eggs can be carried by birds, the freshwater insects all have flying adult stages, land snails or their young can be carried by birds or strong winds, most terrestrial insects can fly and others could be rafted, and many plants have seeds or spores that can be carried by birds or the wind.

There are constraints from the other direction on the late Mesozoic position of Australia. There cannot have been too wide a gap to the northeast, otherwise the continent could not have received propagules from southeast Asia via whatever island stepping stones were available. This route was probably important for early plant life such as the liverwort, conifer, and primitive angiosperm families. However, considering that there were a number of early vertebrate animal imports from South America but none from East Asia until the mid-Tertiary, we can assume that the location of Australia was initially closer to the Antarctic than to the East Indies and that it changed position as plate movement took place.

Chapter 8

ANTARCTICA

> On one hand, Antarctica is the keystone of the mobilist Gondwana fragmentation concept; on the
> other, an ice-free continent in the present position of Antarctica would be in itself a potent connec-
> ting link between the Old and New Worlds.
>
> E. Munroe, *Zoogeography of Insects and Allied Groups,* 1965

Almost every year, new fossil finds are reported from the Antarctic continent. These provide an increasingly better concept of the composition of its Mesozoic and Tertiary fauna and flora. Woodburne and Zinsmeister (1984) have announced the important discovery of the remains of the first land mammal from Antarctica. These were found to represent a new genus and species belonging to the marsupial family Polydolopidae. The fossils came from Seymour Island near the tip of the Antarctic Peninsula. Their age was determined to be late Eocene or about 40 Ma.

In South America, the polydolopid marsupials are most abundant in the fossil record between 50 and 55 Ma. The Antarctic specimens most closely resemble taxa that lived in South America about 50 Ma ago. For these reasons, Woodburne and Zinsmeister (1984) suggested that the Seymour Island animals probably migrated from South America about 10 Ma earlier than indicated by the age of the fossils. These authors also suggested that the marsupial migration was probably the result of a waif dispersal with the Antarctic Peninsula acting as an overland route.

The suggestion that the Antarctic fossils may indicate that a marsupial dispersal from South America took place about 50 Ma ago, corresponds well with the conclusions of Kirsch (1977) who carried out a serological study of marsupial relationships. He found that the Australian taxa were all closely related indicating that they had all evolved in Australia possibly from a very few immigrant forms. He suggested further that all the events of marsupial evolution in Australia had taken place in the last 50 – 60 Ma. This raises the possibility that marsupials may have succeeded in reaching Australia shortly after they had invaded Antarctica.

There has also been a significant fossil discovery in Australia that has clarified the South American – Antarctic – Australian relationships of the early Mesozoic (Lower Triassic). A mammal-like reptile, identified as the genus *Kannemeyeria,* has been reported from Australia (Thulborn, 1983). This genus was previously known from the *Cynognathus* Zone of the Puesto Viejo Formation of South America. In his review paper on the Mesozoic vertebrates of Antarctica, Colbert (1982) emphasized that the Lower Triassic amphibians and reptiles were closely related to fossils previously found in Africa. However, since Africa and South America were broadly joined at the time, the immigration to Antarctica could have taken place from South America.

The early Mesozoic fossils of Antarctica (Colbert, 1982) indicate that it probably provided a suitable habitat for many elements of the worldwide, interrelated vertebrate fauna that existed in the Triassic (Cox, 1974; Charig, 1979; Olson, 1979). In addition, Antarctica served as a passageway allowing some of those species to reach Australia. The relationship to New Zealand has not been as clear-cut but the recent find of early, late Cretaceous dinosaur remains in New Zealand (Molnar, 1981) may mean that a connection or a close proximity to those islands existed at that time. Other possibilities are that the dinosaur was an aquatic creature that could have negotiated a saltwater barrier or that the invasion may have taken place much earlier with the late Cretaceous fossil representing a population that had been isolated on New Zealand for several millions of years. It is likely that the one other old reptile (*Sphenodon*) and the frog genus (*Leiopelma*) reached New Zealand in the Triassic or Jurassic (p. 61).

There is no doubt that, during the Cretaceous and early Tertiary, Antarctica supported a diverse flora that existed under equable temperature conditions (Axelrod, 1984). The early Cretaceous flora described from Victoria, southeastern Australia (Douglas and Williams, 1982) was probably identical to that found on Antarctica itself. It consisted of a variety of pteridosperms, Cycadales, Bennettitales, Ginkoales, Confirales, and Angiospermae. By the early Tertiary, the Antarctic forests became dominated by broad-leaved, evergreen dicotyledons and evergreen conifers. The species appear to be the types that are adapted to a high rainfall distributed throughout the year and low to moderate ranges of temperatures (Axelrod, 1984).

When one adds such fossil evidence to the repeated pattern of contemporary relationships linking the Antipodes to southern South America, the central role of the Antarctic continent in southern hemisphere biogeography becomes obvious. As has been noted in the accounts dealing with Australia and New Zealand, the older biota demonstrates strong ties to South America, ones that must have been continuous through Antarctica. Such relationships involve a broad spectrum of the fauna and flora. Examples are vertebrates such as dinosaurs, mammal-like reptiles, frogs, turtles, and ratite birds; aquatic invertebrates such as the chironomid midges, mayflies, stoneflies, blackflies, and crayfish; terrestrial invertebrates including families of spiders, beetles, robber flies, and bees; and plants such as lichens, bryophytes, ferns, conifers, and angiosperms.

In their paper, Woodburne and Zinsmeister (1984) presented a summary of the late Cretaceous/early Cenozoic paleogeography of the southern hemisphere. They believed that the evidence favors a relatively continuous, overland dispersal route from the tip of South America to the Antarctic Peninsula to the Antarctic continent and thence to Australia and New Zealand. Comparison of shallow-water molluscan faunas from opposite sides of the hypothetical land bridge appear to substantiate the existence of such a barrier. Although magnetic anomaly data appear to indicate that a rift zone began to open between Australia and Antarctica in the late Jurassic, and that deep-sea conditions extended along most of the rift by 80 Ma (Cande and Mutter, 1982), there may not have been a complete deep-water separation until the Eocene (Woodburne and Zinsmeister, 1984).

Although differences in the shallow-water marine faunas on each side of the proposed land bridge might substantiate the presence of such a structure, there may be another explanation. In late Cretaceous/early Tertiary times, the circum-polar West Wind Drift current had not yet developed (Zinsmeister, 1982). But polar sea surface temperatures had declined to the extent that a distinctive warm-temperate fauna had developed in those areas (Briggs, 1974). Since the Pacific Ocean side of the hypothetical bridge was, for the most part, in lower latitudes, that area may have remained tropical while the opposite side became warm-temperate. The presence of such a temperature barrier alone, without the existence of a land bridge, could account for considerable faunal differences.

While a Cretaceous/early Tertiary overland dispersal route may be helpful in solving some distributional problems, it creates others. In South America, there existed a diverse assemblage of early placental mammals contemporaneous with the marsupials. Yet none of the former succeeded in reaching Australia or New Zealand. At the same time, South America also possessed rich amphibian, reptilian, and freshwater fish faunas. None of the fishes and very few of the reptiles and amphibians got across. Woodburne and Zinsmeister (1984) suggested that the Antarctic Peninsula may have filtered out the placental mammals. Such a filter, to be so effective over a long period of time, must have included sea passages of considerable magnitude. We need to keep in mind that such places as Madagascar and the West Indies have, over the years, been successfully invaded by a variety of small mammals, amphibians, reptiles, plants, insects, etc., and that such invasions have taken place by means of waif dispersal across substantial oceanic barriers. It appears more likely that the South American – Antipodes migrations took place across a widely separated archipelago rather than a "relatively continuous" overland route.

In reflecting on the former biogeographic influence of Antarctica, it is important to consider that the continent was probably more than just a connecting link among southern hemisphere land masses. Its size was such that considerable evolutionary activity could have occurred and had its effect on the peripheral land areas. In 1853, the eminent botanist J.D. Hooker speculated on the possibility that the plants of the Southern Ocean were the remains of a flora that was once spread over a larger and more continuous tract of land than now exists in that ocean. This prescient observation was made more than 100 years before the modern theory of plate tectonics.

Chapter 9

SOUTH AMERICA

In the 1930's and 1940's it [continental drift] was not taken seriously by most zoogeographers, not so much because the geophysicists opposed it unanimously, but rather because drift was invoked to explain *very recent* distribution patterns, patterns that must have been due to faunal movements in the later Tertiary and Quaternary.

Ernst Mayr, *Evolution and the Diversity of Life,* 1976

The present biogeographical relationships of South America can, when analyzed, tell us a good deal about the past history of that continent. At various times, it has exchanged organisms with Africa, North America, and Antarctica. Each of these exchanges has had a lasting effect so that the modern South American biota has had three major external sources. In turn, the tropics of South America have evidently served as a center of evolutionary origin for many groups that subsequently enriched other parts of the world.

In regard to the early Mesozoic history, the coefficients of similarity at the family level published by Cox (1974) show that the Triassic vertebrate fauna of South America was closest to that of Africa and Europe, rather than North America. At that time, South America shared 56% of its families with North America but 74% with Africa and Europe. This may indicate that direct communication with North America was not easy and that the principal north-south exchange route was via Africa and Europe. Cox noted that the early terrestrial tetrapods apparently evolved first in the Euramerican area then migrated to the other continents.

The modern herpetofauna of South and Middle America (the Neotropical Region) is extraordinarily diverse. More than 2200 species, in more than 300 genera in 37 families are currently recognized (Duellman, 1979). As noted (p. 61), the earliest migrations from South America southward appear to have involved a rhyncocephalian reptile (*Sphenodon*) and a leiopelmatid frog (*Leiopelma*). It is significant that they were able to migrate, presumably via Antarctica, to New Zealand but not to Australia. Early northward relationships are indicated by the frog family Ascaphidae which is known from an early Jurassic fossil from Argentina and a living genus (*Ascaphus*) in northwestern North America (Estes and Reig, 1973). A second, southward herpetofaunal export phase occurred, apparently much later, when it became possible for a few South American forms to reach Australia, also presumably via Antarctica. This phase included the frog families Leptodactylidae and Hylidae, the freshwater turtle families Meiolaniidae and Chelidae, and a xiphodont crocodile (Tyler, 1979). None of these latter groups reached New Zealand.

To the east, the herpetofaunal relationships with Africa have been reviewed by Laurent (1979) and are discussed in detail under the African account (pp. 101 – 104). Despite several indications of relationship to Africa (direct and indirect), it should be noted that the rich amphibian fauna of South America also contains four frog families (Brachycephalidae, Rhinodermatidae, Dendrobatidae, Pseudidae) and two caecilian families (Rhinatrematidae, Typhlonectidae) that are not found in Africa or anywhere else.

Although the freshwater turtle family Pelomedusidae has a South American – African – Madagascar range at present, it has been found in the Cretaceous of Europe and North America and in the Eocene of Asia. It also occurs in the Cretaceous of South America and Africa (Baez and de Gasparini, 1979). The oldest fossil is apparently from the Lower Cretaceous of Africa. Some of these early pelomedusid turtles may have also existed in the marine environment (Wood, 1976). One living genus (*Podocnemis*) is found in South America and Madagascar but not Africa. The other freshwater turtle families in South America either have northern (4) or Australian (2) relationships (Duellman, 1979).

While the living crocodilians of South America do not illustrate any African relationships, the Cretaceous pholidosaurid genus *Sarcosuchus* was apparently common to Brazil and West Africa (Laurent, 1979). Among the lizards, only the Gekkonidae and the Scincidae also occur in Africa. The gekkos comprise a large tropicolitan group that probably originated in the Upper Jurassic/Lower Cretaceous (Kluge, 1967). Gekkos are very good at waif dispersal (rafting) and modern genera and species are shared with Africa. The Iguanidae is the largest family of lizards in the New World where it is represented by more than 50 genera. It evidently evolved primarily in South America but probably also existed in Africa since there are two living genera isolated on Madagascar. Its extinction in Africa may have been due to competition from the agamid and chamaeleontid lizards. The iguanids probably migrated to North America in the Paleocene (Estes, 1983). The skinks are predominately an African family with only a small representation in South America.

Amphisbaenians are small, specialized burrowing reptiles that do not seem to be closely related to either lizards or snakes (Goin and Goin, 1971). They are discussed under the African account (p. 103). The slender blind snakes (Leptotyphlopidae) belong to a single widespread genus that is present in both South America and Africa. A western Gondwanaland (South American) origin has been suggested (Duellman, 1979). The blind snakes (Typhlopidae) comprise a small family that is pantropical; it has been called pan-Gondwanian by Laurent (1979). The snake families Boidae, Colubridae, and Viperidae are discussed later (p. 104) but it may be noted here that the viperine subfamily Crotalinae apparently arose in the Oriental Region and invaded the New World via the Bering Land Bridge. The three South American crotaline genera represent invasions from North America.

The presence of oceanic passages through the island archipelago that existed, in various configurations, between South and North America during the Cenozoic, comprised a significant barrier to migration by terrestrial and freshwater animals. However, as a group, the herpetofauna contained many animals that were adept at waif dispersal. In the early Tertiary, at least five amphibian groups moved northward from South America (Duellman, 1979). Toward the end of the Tertiary, the plethodontid salamanders invaded South America from the north (Wake and Lynch, 1976). The reptilian exchanges were even more extensive. Duellman (1979) listed eight families that were apparently exported northward and four that entered South America for the first time. Once the isthmian link was established in the late Pliocene or early Pleistocene, a migratory flood resulted with representatives of 23 families moving north and 18 moving south.

The Neotropical Region, including South and Central America, has a relatively rich and distinctive bird fauna. It contains about one third of all known species of birds. Of the 86 families, 31 are endemic. This diverse bird fauna is comprised, to a large extent, of relatively primitive families (Welty, 1979). There are such peculiar groups as the Tinamidae (Tinamous), Rheidae (rheas), Cracidae (curassows), Aramidae (limpkins), Psophidae (trumpeters), Ceriamidae (seriemas), Steatornithidae (oilbirds), and nine families of the more primitive passeriforms. These give the bird fauna of the Neotropics a decidedly primitive cast.

Mayr (1976) called attention to the surprising lack of relationship between the bird faunas of South America and Africa. He noted that hardly any other two bird faunas are more different and that there was almost no faunal sharing between the two continents. Mayr considered only five families of freshwater birds and three families of land birds to be circumtropical. The latter are the barbets (Capitonidae), trogons (Trogonidae), and the parrots (Psittacidae). These families were probably able to migrate between South or Middle America and Africa when the continents were closer to one another. The trogons probably originated in the New World while the barbets and parrots came from the Old World.

Mayr (1946) noted that the South American bird fauna had been enriched by a relatively recent influx of land bird families from North America including the jays (Corvidae), wrens (Troglodytidae), thrushes (Turdidae), vireos (Vireonidae), honeycreepers (Coerebidae), wood warblers (Parulidae), blackbirds (Icteridae), tanangers (Thraupidae), cardinals (Pyrrhuloxiinae), and finches (Fringillidae). In contrast, only two families of South American origin, the hummingbirds (Trochilidae) and the tyrant flycatchers (Tyrannidae), have successfully invaded North America. Most of these exchanges probably took place after the establishment of the isthmian link.

The history of mammals in South America is interesting and, to some extent, controversial. Cretaceous marsupial fossils are now known from both South and North America (Archibald and Clemens, 1984). Marsupials reached Europe by the Eocene and, by the early Oligocene, were present in north Africa (Brown and Simons, 1984) and in central Asia (Gabunia and Shevyreva, 1985). The African and Asian fossils are evidently related to the European – American forms so probably came from Europe. By Miocene times, all of the northern hemisphere marsupials became extinct. In South America the order underwent a significant evolutionary radiation. As noted (p. 81), a serological study by Kirsch (1977) showed greater differences among the various South American marsupial groups than among those in Australia, indicating a longer period of evolution in the former.

It has been commonly assumed that marsupials arose in North America, then dispersed to South America and other localities (Colbert, 1980) but the discovery of late Cretaceous fossils in South America, and the evidence of their old and extensive radiation on that continent, make Archer's (1984) suggestion of a South American origin a real possibility. There appears to be no good reason why the marsupials (and their sister group the placentals) could not have evolved from their eupantothere ancestors in South America. From there, they could have reached, by means of waif dispersal through Central America, North America in the late Cretaceous. With the drying up of the Mid-Continental Sea at the end of the

Cretaceous, they could have expanded to Europe. From there, they could have crossed relatively narrow sea barriers to reach north Africa and central Asia. This possible distributional sequence is illustrated on a Paleocene map (Fig. 11).

The placental order Edentata is poorly represented in the fossil record. It possibly appeared in South America during the late Cretaceous or early Paleocene but an edentate has been reported from the Paleocene of China (Ding, 1979). During the Tertiary, the South American edentates evolved as armadillos and glyptodonts on one hand and as ground sloths, tree sloths, and anteaters on the other. The caviomorph rodents comprise a large group of about 10 living and several fossil families. Fossils have been found in Texas and central Mexico (Colbert, 1980) and it is thought that the group may have reached South America by rafting in the late Eocene or early Oligocene. The long isolation in South America resulted in the evolution of many peculiar families such as the guinea pigs, capybaras, fake pacas, agoutis, pacas, chinchillas, and porcupines.

The hooved mammals called ungulates probably all arose from members of a primitive order called the Condylarthra. Condylarths are known from the late Cretaceous rock of both South and North America and from the early Cenozoic of the Americas, Europe, and Asia (Simpson, 1980). The condylarths of South America gave rise to a peculiar ungulate fauna that developed in almost complete isolation. The great majority of the South American ungulates belonged to the

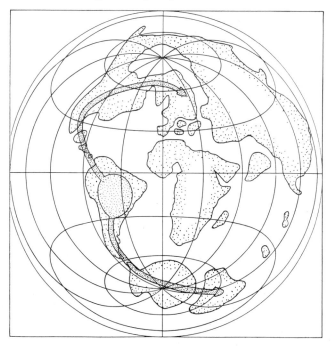

Fig. 11. Possible dispersal of marsupials from a center of origin in South America. Movement to North America would have taken place in the late Cretaceous. Southward dispersal to Antarctica and Australia may have taken place in the Paleocene (the Period of this map). Later, marsupials from Europe reached north Africa and Central Asia.

Order Notoungulata. In its entirety, this old diverse fauna consisted of nine orders and 22 families, most of them bizarre forms completely unlike the hooved mammals of other continents.

The New World monkeys are considered to be the most primitive of the three higher primate groups, the other two being the Old World monkeys and the great apes together with man. The oldest primate known from South America is *Branisella* from the early Oligocene of Bolivia (Simpson, 1980). It is possible that the tarsioid family Omomyidae, fossils of which have been found in Wyoming, represents the stem group from which the New World monkeys evolved (Szalay, 1975; McKenna, 1980). The omomyids may have undergone a waif dispersal to South America in the Eocene to begin the evolution of the New World monkeys (families Cebidae and Atelidae).

However, the history of the New World monkeys is still, despite a large amount of research on primate relationships, highly controversial. Another view is that they are very closely related to the Old World group and, therefore, must have migrated across the Atlantic directly from Africa (Hoffstetter, 1980; Lavocat, 1980). But, by Eocene times, when the migration supposedly took place, the Atlantic Ocean was large enough (Map 7) to present a formidable barrier (van Andel et al., 1977). At this date, it is fair to say that no unequivocal example of direct relationship between South America and Africa has been demonstrated. It is probable that all of the mammalian groups just discussed (marsupials, edentates, caviomorph rodents, ungulates, and New World monkeys) developed in isolation in South America throughout most of the Tertiary after originating in South America or having been derived from North American ancestors.

In the Cretaceous and early Tertiary, there were relatively few exchanges of mammalian stocks between North and South America. As noted, marsupials and even some early placentals may have migrated northward in the late Cretaceous. Some early Tertiary notoungulate fossils have turned up in North America and Asia and a Paleocene edentate has shown up in China (Savage and Russell, 1983). In the late Miocene there was a minor exchange with the raccoon (Procyonidae) reaching South America and two genera of small ground sloths going the opposite direction (Webb, 1985a).

With the advent of the complete land connection between the Americas, the great mammalian interchange got underway reaching its peak in the Pleistocene (Simpson, 1980; Marshall et al., 1982; Webb, 1985a). Representatives of 15 families of North American mammals entered South America and seven families migrated in the opposite direction. In South America, the effect was catastrophic and resulted in the extinction of the unique notoungulates, litopterns, and marsupial carnivores. Comparatively, the invasion of Central and North America by South American mammals was not nearly so successful. Only three migrants managed to survive north of Mexico, an opossum, an armadillo, and a porcupine.

In the freshwater environment, it has already been noted (p. 71) that certain South American genera of the mussel family Hyriidae belong to a subfamily that is related to two other subfamilies found only in Australia and New Zealand. Parodiz and Bonetto (1962) described two related families, the Mycetopodidae of South America and the Mutelidae of Africa, that are related by possession of a unique lasidium larva. The general distribution of the freshwater snail family Potamiopsidae (Davis, 1979) shows a high diversity of advanced genera in southeast Asia with the three most primitive genera in South America, South Africa, and Australia respectively. Other molluscan relationships have been discussed by Jaeckel (1968).

As noted (p. 71), the freshwater crayfish family Parastacidae (Bishop, 1967) includes a South American genus (*Parastacus*) with other genera occurring in Australia, New Zealand, New Guinea, Aru Islands, and Madagascar. The freshwater brachyuran family Pomatomidae has its evident center of origin in southeast Asia (Bishop, 1967) but one subfamily occurs in South America and the other is widespread from Africa and southern Europe to Australia.

The waterstriders belonging to the family Gerridae have been revised by Calabrese (1980). She found 18 Neotropical genera with 11 considered to be endemic. Of the non-endemic genera, two are worldwide, two are shared with Africa (one continuing to the Oriental Region), and three with North America; the latter probably represents invasions from South America. The parasitic (mainly on freshwater fishes) nematode family Camallanidae (Stromberg and Crites, 1974) is represented in South America by a depauperate and primitive fauna. The South American fauna probably came from Africa in the Mesozoic with the invasion of ostariophysan fishes.

The distribution of Neotropical freshwater insects was reviewed by Illies (1968). Paying particular attention to the stoneflies (Plecoptera), he distinguished two major groups, one consisting of paleantarctic elements which also occur in Australia and New Zealand and a Nearctic group that had invaded from the north. The relationship to Africa was considered to be relatively weak.

The archaic mayfly (Ephemeroptera) family Leptophlebiidae has a predominately southern, cool temperate distribution. Recent systematic work on the South American genera (Pescador and Peters, 1980) indicated that they are most closely related to genera from Australia followed by genera from New Zealand, Africa, and Madagascar, in that order. Edmunds (1972) felt that the general pattern of mayfly distribution gave overwhelming evidence that the closest southern relationships were between southern South America and Australia. Edmunds (1982) also noted that when the isthmian connection was completed, 21 South American mayfly genera migrated northward but only one genus moved in the opposite direction. He noted, furthermore, that some lineages were present in South America and Madagascar but absent in Africa.

In the caddisflies (Trichoptera), Ross (1967) called attention to an old relationship among South America, Australia, and New Zealand. Evans (1959) summarized the information then available on other aquatic insect groups: included were the dragonflies (Odonata), dobson flies (Megaloptera), the Blepharioceridae and the Simuliidae (Diptera), scorpion flies (Mecoptera), and two neuropteran subfamilies. Many genera in these groups indicated close South American – Australian relationships. South American – African relationships have been noted among the dragonflies (Pinkey, 1978), mayflies (Edmunds, 1972), and for the caddisflies, waterbeetles (Torridincolidae), and a freshwater snail (Planorbiidae) by Harrison (1978). In general, the relationship to Africa among the aquatic insect groups seems to be considerably weaker than to Australia.

The evidence presented by Grekoff and Krommelbein (1967) from the fossils of Upper Jurassic South American and African freshwater ostracods is of considerable interest. They indicated that specimens collected from northeastern Brazil and from West Africa must have been parts of one original assemblage from a single basin. Besch (1968) called attention to the zoogeographic value of the water mites (Hydrachnellae) with the genus *Australiobates* being found in southern South America, South Africa, and Australia and with the genus *Lundbladobates* occurring in South America and Australia only.

The ostariophysan fishes comprise a group of related orders and families that are characterized by the possession of a Weberian apparatus for the transmission of sound impulses from the swimbladder to the inner ear. This evolutionary innovation has enabled the ostariophysans to become very successful, in fact, to dominate the freshwater habitats to which they have been able to gain access. They are considered to be primary freshwater fishes (Myers, 1938) and have evidently been confined to freshwaters almost throughout their history. Of the major groups that comprise the ostariophysans, the characoids (Suborder Characoidea) are generally considered to be the most primitive. In the Neotropics, they are represented by 15 families but there are only four families in the Ethiopian Region. Only one family, the Characidae, is shared by the two continents but the genera in each place probably belong to different subfamilies. The characoids are now restricted to the Neotropical and Ethiopian Regions but fossils have been identified from Paleocene and Eocene deposits in France and England (Patterson, 1975, 1981a).

In the ostariophysan order Siluriformes (catfishes), all 13 families that exist in the Neotropical Region are endemic. Of the six families that are found in the Ethiopian Region, three are endemic and the other three are shared with the Oriental Region. The earliest fossil siluroids occur in the Cretaceous of Bolivia (Wenz, 1969). The third major ostariophysan group is the cyprinoids (Suborder Cyprinoidea). They are not found in South America.

It has been suggested that the early characoids, followed closely by the first siluriform fishes, originated in the Oriental Region in the Upper Jurassic (Briggs, 1979). From tropical Asia, these groups probably spread to Europe via an occasional trans-Turgai connection. Once in Europe, they could have migrated around the west end of the still incomplete Tethys Sea and entered Africa (Fig. 12). Their entry to South America was probably from Africa by means of a peninsular connection that was in existence in the Upper Jurassic. The cyprinoid fishes, the most advanced ostariophysan group, evidently did not enter Africa until the Miocene, too late to continue on to South America.

There are other groups of primary freshwater fishes but, compared to Africa, South America has very few. These are a lungfish (Lepidosirenidae) which is related to the African Protopteridae, two genera of Osteoglossidae, and two genera of Nandidae. The latter two families both also occur in Africa. Darlington (1957) suggested that, due to its limitation and imbalance, the South American primary freshwater fish fauna is a derived one, descended from a few immigrants that somehow reached South America from Africa. More recent research has indicated that this suggestion is probably correct.

The order Cyprinodontiformes comprises a large group of about 900 teleostean fishes commonly known as killifishes, topminnows, or toothcarps. Traditionally, the order includes five families, the oviparous Cyprinodontidae, which is widespread in the New and Old Worlds, and the New World viviparous families Anablepidae, Goodeidae, Jenynsiidae, and Poeciliidae. As a whole, the group is considered to belong to the secondary freshwater fish category (Myers, 1938) since many of the species live in brackish water and some can tolerate very high salinities.

The phylogeny and distribution of the cyprinodontiform fishes has been examined by Parenti (1981). She recognized two suborders, the Aplocheiloidei and the Cyprinodontoidei. The former was divided into groups of eleven New World and five Old World genera. The New World group is essentially confined to the Neotropics and the West Indies while that of the Old World is found in tropical Africa, Madagascar and southeast Asia. The Cyprinodontoidei include the four New World viviparous groups plus many oviparous genera from both the New and Old Worlds.

According to Parenti (1981), links between New and Old World cyprinodontoids occurred at three different levels in the evolutionary development of the suborder. The most advanced of these links is made by the recognition of the genus *Orestias,* which is endemic to the high-altitude lakes of the Andes, as being most closely related to the Anatolian genus *Koswigichthys.* The tribe (Orestini) into which these

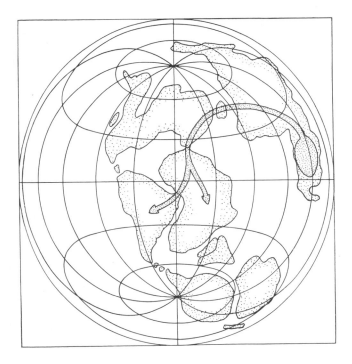

Fig. 12. Suggested late Jurassic dispersal route of primitive ostariophysan freshwater fishes (characins and catfishes) from a possible center of origin in southeast Asia.

genera are placed is considered, together with its sister group (the Cyprinodontini), to be the most advanced of all the major groups of cyprinodontiform fishes.

While the general distribution of the killifishes indicates that they can transgress minor saltwater barriers, having reached some places (such as the West Indies and Madagascar) where primary freshwater fishes do not occur, they are stopped by wider or more numerous stretches of saltwater and have not penetrated such places as Australia or New Zealand. Considering these limitations, passage between the tropics of the New and Old Worlds was probably not possible except when South America and Africa were joined or separated by only a narrow oceanic barrier.

Parenti (1981) has given the late Triassic as an estimate of the minimum age for the order since by that time the Laurasian and Gondwanian land masses had begun their separation. However, the oldest described cyprinodontiform fossil is *Prolebias,* an Anatolian cyprinodontine from the Oligocene of Europe. The aplocheilids of the Old World have generally been considered to be the most primitive members of the order and Myers (1958) noted that *Aplocheilus* presented the largest number of basic characters that had become specialized or lost in other members of the family.

Instead of considering the cyprinodontiform fishes to have been a Pangaean group with an origin in the late Triassic (which would place them in a pre-teleostean fauna) an alternative would be to hypothesize a later origin, perhaps early Cretaceous. By the latter time, South America had probably separated from Africa but the two continents were still in close proximity. A primitive aplocheiloid with the ability to live under euryhaline conditions could have invaded Africa and gradually spread to other parts of the Old World.

Although the classification published by Parenti (1981) differs considerably from early views, it is still apparent that generic diversity is highest in the New World (28), next highest in Africa (11), lower in the Near East (4), and least in southeast Asia (1). Furthermore, all of the viviparous groups with their many reproductive specializations are endemic to the New World. It has been suggested, therefore, that the cyprinodontiform fishes originated in the New World, probably in the area from Mexico to South America, and succeeded in reaching the Old World via Africa early in their evolutionary history (Briggs, 1984a). By the late Cretaceous, continental drift had proceeded to the extent that it was probably no longer possible for cyprinodontiform fishes to migrate between South America and Africa. Their probable distributional routes are indicated on an early Cretaceous map (Fig. 13).

There are some general patterns of interest among the terrestrial invertebrates. Peake (1978), in discussing the zoogeography of the Stylommatophora, the dominant group of land snails, commented that the present relationships between South America and Africa depended largely on taxa with worldwide distributions and that there was a notable absence in the fossil record of a clear affinity between the two continents. He concluded, "For the majority of families it can be argued that the recent distributions reflect the late Mesozoic or Tertiary arrangement of continents, subsequent to the fragmentation of the land masses, far more satisfactorily than any simple division into Gondwanaland or Laurasia."

In discussing the relationships of the terrestrial insects of Australia (p. 74), it was noted that there were many cool-temperate groups that indicated a close affinity to the southern part of South America. Examples are the Peloridiidae (Evans, 1959), the archaic dipteran fauna (Paramonov, 1959), aphids that infest *Nothofagus* (Schlinger, 1974), older elements of several major beetle families (Howden and Cooper, 1977; Thiele, 1977; Howden, 1981), and archaic groups of Lepidoptera (Tindale, 1981). Also, Darlington (1965) called attention to such a relationship in a family of primitive bugs (Homopteridae) and in four different tribes of carabid beetles. Similar ties to Australia (sometimes including New Zealand) were found in the primitive flatbugs (Usinger and Matsuda, 1959), the robber flies (Hull, 1962), the fleas (Traub, 1972), and the Aeolothripidae (Mound, 1972).

Some of the primitive South American insect groups also show ties to the southern part of Africa but the relationship is not as close as to Australia. Among the camel crickets (Orthoptera), the subfamily Macropathinae has species in South America, South Africa, southeastern Australia, and New Zealand (Rentz, 1978). In the beetle family Lucanidae, the subfamily Lampriminae has a current southern distribution, existing in South America, South Africa, and Australia; but specimens have also been found in the Baltic Amber of the Eocene/Oligocene (Endrody-Younga, 1978). Reichardt (1979) has noted that the primitive beetle tribe Ctenostomini is restricted to the Neotropical Region and Madagascar and that other groups of Coleoptera show a similar pattern. Such groups must have existed in Africa at one time but have become extinct there.

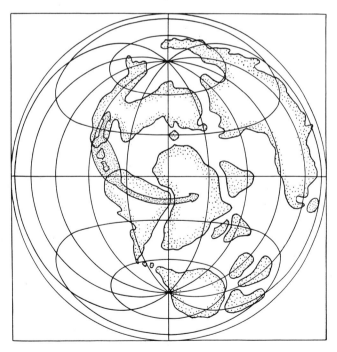

Fig. 13. Possible early Cretaceous dispersal of cyprinodontiform fishes from an evolutionary center in Central America. Later, possibly by the Miocene, these secondary freshwater fishes were able to reach Madagascar and India.

Both South America and Africa have large termite faunas but the major center of origin has probably been in the latter. As noted (p. 74), the Mastotermitidae, considered to be the most primitive family, has a living species in Australia but a related genus has been reported from the Miocene/Pliocene of Brazil (Araujo, 1970). In the Hodotermitidae, the genus *Porotermes* is represented by three species, one each at the southern tips of South America, South Africa, and Australia – Tasmania. The largest family of termites is the Termitidae; its subfamily the Nasutotermitinae probably evolved in the Neotropics at the beginning of the Cretaceous (Krishna, 1970). All but two of the 19 genera are endemic.

The richest bee fauna (315 genera and subgenera) is found in the Neotropics (Michener, 1979). In comparison, the African and Oriental tropics have relatively poor bee faunas. The Nearctic is richer than the Palearctic largely because of the invasion of Neotropical elements. Michener stated that the place of origin might have been the arid interior of West Gondwanaland. There is a notable southern distribution of primitive groups. The Paracolletini occurs in temperate South America, South Africa, and Australia. The primitive family Fedeliidae is limited to the arid parts of Chile and southern Africa.

As was noted in the description of the relationships of the Australian flora (p. 75), many groups of primitive plants are related to those occurring in southern South America. Included are the liverworts (Schuster, 1969), the conifers *Araucaria, Podocarpus,* and cedars (Florin, 1963); also some of the early angiosperms such as the southern beech *Nothofagus*. In South America only, *Nothofagus* is symbiotic with a unigeneric family of angiosperms, the Myzodendraceae (Schuster, 1972). In general, such primitive plants are concentrated in the area stretching from southeastern Asia out to Australia and New Zealand. Those that exist in South America may have migrated from Australasia via Antarctica.

When one looks at the general distribution of the families of angiosperms, the independence of the Neotropical flora becomes clear. No less than 47 families are endemic to that part of the world. When one includes the Bromeliaceae with about 1500 species and the Cactaceae with more than 2000 species, both virtually confined to the New World, it can be seen that the American tropical flora is highly distinctive (Good, 1974). In comparison, tropical Africa has only 17 endemic families and southeast Asia 15. There are about 60 families that occur in all three main tropical regions but only about 12 that are restricted to South America plus Africa.

In temperate South America, there are some 14 endemic families partly or entirely in that area, 10 in southern Africa, 12 in Madagascar, and 19 in Australia (Good, 1974). About a dozen families exist in all three sectors (South America, Africa, Australasia) of the southern hemisphere. In addition, there is a group of nine families that are discontinuously distributed between South America and Australasia, the three largest being the Epacridaceae, Goodeniaceae (exempting two pan-tropical species), and the Stylidiaceae.

At the generic level, one can find many examples of discontinuous distribution in the south temperate regions. There are about 50 such genera in South America, every one occurring in some part of Australasia but only three (*Acaena, Gunnera, Tetragonia*) are found also in southern Africa (Good, 1974). There seems to be no genus recorded only from temperate South America and southern Africa. Most of the South American – Australasian genera are small and fairly evenly balanced in species between the two regions but some (*Fuchsia, Gaultheria, Pernettia*) have a strong preponderance of species in South America. Aside from the very primitive groups already mentioned, very few (*Colobanthus, Oreomyrrhis*) are best developed in Australasia.

While there are many genera (about 250) that are confined to the tropics and occur in all three major tropical areas, there are also many tropical genera that demonstrate more restricted distributions; Good (1974) listed 97 that are shared between the Neotropical and Ethiopian (including Madagascar) Regions only. In comparison, the total number of genera unknown outside tropical America appears to approach 3000. About 38 genera are shared only between the Neotropics and the Oriental Region. A number of species are shared between the tropics of America and Africa; examples are provided by Good (1974). Some 32 species are shared between the cool-temperate parts of South America and Australasia (Moore, 1972).

Raven and Axelrod (1974) and Axelrod and Raven (1978) envisioned a rather late (mid-Cretaceous) final separation of South America and Africa. They suggested that the two continents shared a common angiosperm flora into Eocene time and that the ties between the floras are now chiefly at the level of families. However, as was noted above, the family level relationships show that South America has a highly peculiar flora, one that must have largely evolved after its separation from Africa. Both Smith (1973) and Thorne (1973) emphasized the lack of relationship between the two floras. Angiosperms probably first began to arise in the Lower Cretaceous (Brenner, 1976). The present relationships of the flora of South America and Africa appear to be consistent with the hypothesis that the two continents were separated by that time.

Although some studies of plate movements (i.e. Dietz and Holden, 1970; Barron et al., 1981) indicated that South America and Africa were still broadly joined at the beginning of the Cretaceous, data from the marine environment seems to correlate better with information based on the relationships of the terrestrial and freshwater biota. Kauffman (1973) recognized a distinct South Atlantic subprovince for the bivalve Mollusca beginning in the Albian (about 108 Ma) of the Lower Cretaceous. Rawson (1981) found that endemic ammonite taxa were present in the same area at about the same time. Even more important, the study of the depositional history of the South Atlantic by van Andel et al. (1977) indicated the presence of isolated, narrow marine basins between South America and Africa for the earliest part of the Cretaceous (about 140 – 150 Ma). Tarling (1980) demonstrated the presence of extensive evaporite deposits between the continents during Aptian times (110 Ma). Thus, a series of narrow marine plus evaporite basins probably constituted a significant biogeographic barrier during the early Cretaceous even though the continents were still close to one another.

SOUTH AMERICA SUMMARY

More than any other continent, South America is a land of biological contrasts. Its cool-temperate, southern tip serves as a refuge for many primitive animals and plants that were once widespread. In comparison, the South American tropics contain a diverse assemblage of modern organisms, many of which have successfully invaded other continents. Especially striking is the enormous number of species in such groups as the birds (about one-third of all known species), frogs (a majority of the world's species), and bromeliad plants (most of the known species).

There was a time in the early Mesozoic (the Triassic) when a global warm climate and a close proximity of the continents made it possible for a similar reptile fauna to exist on all the major land masses. South America evidently derived its early tetrapod fauna from a Euramerican distributional center either directly or via Africa. This Triassic fauna then managed to enter Antarctica and thence Australia. A small, lizard-like, rhynchocephalian reptile made it to New Zealand and, probably much later, a dinosaur showed up in that country.

The various groups of the South American biota indicate, at different times, relationships to North America, Africa, and the Australian – New Zealand area. In the herpetofauna, the anurans (frogs) are very diverse and have undergone much of their evolution in South America. The most primitive living genera are found in New Zealand and in western North America but they are related to two fossil genera described from the Jurassic of Patagonia. Some groups apparently originated in South America in the late Mesozoic or early Tertiary and migrated, presumably via Antarctica, to Australia. Examples are two frog families (Leptodactylidae and Hylidae), two freshwater turtle families (Meiolaniidae and Chelidae), and a xiphodont crocodile.

There are also early herpetofaunal relationships to Africa. The caecilians (Gymnophiona) are most diverse in South America but are well represented in Africa. Despite the presence of several amphibian families that are shared with Africa, there are still two caecilian and four frog families that are endemic to South America. Among the lizards, only the Gekkonidae and the Scincidae are shared with continental Africa. In the herpetofauna in general, however, it is interesting to find some special relationships to Madagascar. The freshwater turtle genus *Podocnemis* occurs in both South America and Madagascar but not in Africa. A similar pattern is found in the iguanid lizards, a New World family with two genera isolated on Madagascar, and the snake subfamily Boinae which also has two genera on Madagascar. In all three cases, the groups must have been present in Africa but were probably eliminated as more advanced reptiles entered or developed in Africa. Other reptile groups that demonstrate a South American – African relationship are the amphisbaenians, Leptotyphlopidae, Typhlopidae, Colubridae, and Viperidae.

The relationship of the herpetofauna of South America to that of North and Middle America (and the West Indies) has been summarized by Duellman (1979). In the early Tertiary, at least five amphibian family groups dispersed from south to north. Toward the end of the Tertiary, the plethodontid salamanders invaded South America from the north. Eight families of reptiles were exported northward and four others were imported from the north. Once the isthmian link was established in the late Pliocene or early Pleistocene, a migratory flood took place with representatives of 23 families moving north and 18 moving south.

The bird fauna of South America, compared to that of the other continents, is both diverse, and primitive. It also displays a very high degree of endemicity with

31 of the 86 families being found nowhere else. Two of its large families, the parrots (Psittacidae) and the pigeons (Columbidae and allies) are best developed in the Australian region and may have entered South America via Antarctica or southern Africa. Many primitive bird families have managed to survive in South America but that continent has also served as a center of origin for others. Two such families, the hummingbirds (Trochilidae) and the tyrant flycatchers (Tyrannidae) have successfully invaded North America.

The South American bird fauna was enriched, probably when the isthmian link was completed, by an influx of land bird families from North America including the jays (Corvidae), wrens (Troglodytidae), thrushes (Turdidae), vireos (Vireonidae), honeycreepers (Coerebidae), wood warblers (Parulidae), blackbirds (Icteridae), tanagers (Thraupidae), cardinals (Pyrrhuloxiinae), and finches (Fringillidae). Mayr (1946) called attention to the surprising lack of relationship between the bird faunas of South America and Africa.

The history of mammals in South America is interesting and still controversial. The marsupials either originated in South America or managed to invade from the north, probably in the late Cretaceous, and then underwent a significant evolutionary radiation. Possibly about the same time, the edentates and the primitive condylarth ungulates arrived or developed in situ. The caviomorph rodents probably came a little later (Eocene or early Oligocene). It has been suggested that the tarsioid family Omomyidae, a primitive primate group, came to South America in the Eocene. During the Tertiary, there developed from these few predecessors, an incredibly rich mammalian fauna.

South America evidently exported a few of its marsupials to Australia via Antarctica, an event which probably took place 50 – 60 million years ago. Unlike the herpetofauna, there is not much evidence for the dispersal of mammalian stocks northward or southward during the latter part of the Tertiary. In the late Pliocene, some small ground sloths went north and the raccoon came south. With the completion of the Panamanian Isthmus, a great interchange got underway with 15 families of North American mammals entering South America and 7 families spreading in the opposite direction. The catastrophic effect in South America, resulting in the extinction of the unique notoungulates, litopterns and marsupial carnivores, has been well documented (Simpson, 1980).

In the freshwater environment, one mussel family is shared with Australia and New Zealand while another is related to an African family. A crayfish family has related genera in the general Australian region and in Madagascar and a gastropod family has primitive genera in South America, South Africa, and Australia. A freshwater brachyuran family is represented by a subfamily in South America and another subfamily in Africa and in southern Europe to Australia. Most of the waterstriders (Gerridae) are endemic but two genera are shared with Africa. A family of parasitic nematodes is represented by a depauperate, primitive fauna in South America that apparently reached there from Africa in the Mesozoic.

The general pattern of mayfly distribution indicates the closest generic level relationships to be with Australia followed by New Zealand, Africa, and Madagascar. Old ties between South America and Australia – New Zealand are demonstrated by the caddisflies, dragonflies, dobson flies, blepharocerids and simuliids (Diptera), scorpion flies, and two neuropteran subfamilies. South American – African relationships were noted among some dragonflies, mayflies, caddisflies, water beetles, and a freshwater snail. These relationships were found to be strongest with Australia and considerably weaker with Africa.

Upper Jurassic freshwater ostracods from northeastern Brazil and West Africa were considered to belong to the same population. One genus of water mites is shared among southern South America, South Africa, and Australia while another occurs only in South America and Australia. Among the fishes, the primary freshwater ostariophysans provide some important information about Mesozoic relationships. The most primitive group is the characoid fishes with 15 families in South America and 4 in Africa. One family is shared between these two continents.

Among the catfishes (Siluriformes), the 13 South American families are all endemic. There are 6 families in Africa with 3 being endemic and the other 3 being shared with the Oriental Region. The most specialized ostariophysan group, the cyprinoids, evidently penetrated Africa in the Miocene but did not reach South America. It was suggested the ostariophysans arose in the Oriental Region in the Upper Jurassic with the two most primitive groups (characins and catfishes) migrating to Africa around the west end of the still incomplete Tethys Sea and then reaching South America by a peninsular connection. In general, the African primary freshwater fish fauna is far more diverse in higher taxa than that of South America. It seems that the Upper Jurassic connection to Africa acted as a filter barrier permitting the entry of only a restricted sample of the African freshwater fauna.

The secondary freshwater fish order Cyprinodontiformes evidently arose in the New World, probably in late Jurassic or early Cretaceous times. A primitive cyprinodontiform species with the ability to live under euryhaline conditions could have reached Africa when that continent was still located close to South America. In the Old World, the order was probably initially confined to Africa from which it spread to Madagascar and the Seychelles. The group is less diverse in southeast Asia possibly indicating that it reached that area more recently.

A considerable amount of scattered information about terrestrial invertebrates is available. In the dominant group of land snails, there was no clear affinity demonstrated between South America and Africa and very few families exhibited southern disjunct distributions. Many of the South American cool-temperate groups of insects show a closer relationship to Australia than to southern Africa. Some groups of beetles that now have a strictly southern hemisphere (gondwanic) distribution are known to have existed in the north since their fossils have been found there. Such examples show that caution must be used in assuming a strictly Gondwanaland history for old insect groups.

The major center of termite evolution has probably been Africa, and South America undoubtedly received much of its early (Mesozoic) fauna from there. The most primitive termite groups are south temperate and the most primitive living species exists in Australia but a related fossil genus has been reported from the Miocene/Pliocene of Brazil. The Neotropics has the world's richest bee fauna. There is a notable southern distribution of the more primitive groups with one tribe being shared among the southern parts of South America, Africa, and Australia.

The plantlife of southern South America contains many primitive groups that demonstrate a close relationship to the flora of Australia – New Zealand and associated islands. This affinity is evident among such groups as the liverworts, primitive conifers such as *Araucaria* and *Podocarpus*, and early angiosperms such as *Nothofagus*.

About 12 higher families are found in all three sectors (South American, African, Australian) of the southern hemisphere and nine others are shared between South America and Australasia. About 50 genera demonstrate the latter connection but only three of them also occur in southern Africa. The modern flora of the tropical parts of South and Central America is highly distinctive with 47 families being endemic. In constrast, Africa has only 17 endemic families. A good many of the angiosperm families (about 60) are circumtropical but only a few (about 12) indicate any special South American – African relationship. About 97 genera are shared between tropical South America and Africa but this number is small compared to the 3000 that are endemic to the Neotropics.

The general picture of a high degree of independence in the Neotropical flora at the family and generic levels is not compatible with the supposition (Axelrod and Raven, 1978) that the ties to Africa are chiefly at the family level but include some tribes and genera. The extensive endemism in the Neotropics at both the family and generic levels indicates that in the mid-Cretaceous South America and Africa were probably farther apart than has been indicated from plate tectonic data. Other groups of the biota shed additional light on this problem.

Information provided by marine fossils from the South Atlantic seems to support the concept of a significant barrier between South America and Africa in the earliest part of the Cretaceous. It appears that isolated marine plus evaporite basins at the beginning of the period were soon followed by the establishment of a continuous, open ocean environment which was present by about 108 Ma.

Chapter 10

AFRICA

> I am a firm believer that without speculation there is no good and original observation.
>
> Charles Darwin in his letter to Alfred Russel Wallace, Dec. 22, 1857

The historic relationships of Africa to the other continents are quite different than those of South America. Although Africa was probably joined to South America in the Triassic and maintained some connection until the Upper Jurassic, the relationship of the two continents became increasingly distant from then on. The flora and fauna of South America developed in almost complete isolation throughout the Tertiary. Africa, on the other hand, had intermittent contact with Eurasia in the Tertiary so that its biological history is more complicated. The relationships of Africa to India and Madagascar, while not well understood, offer some interesting insights into the African past.

The relationships of the modern herpetofaunas of Africa and South America have been reviewed by Laurent (1979). The caecilians (Order Gymnophiona) are worm-like amphibians that may be related to the early, scaled amphibians of the Carboniferous. They are most diverse (three families) in South America. The only known fossil caecilian (*Apodops*) is from the Paleocene of southern Brazil. Estes and Wake (1972) observed that the African genus *Geotrypetes* was probably closely related to the fossil South American form. The family Caeciliidae is found on both continents as well as the Seychelles Islands and southeast Asia.

The modern salamanders (Urodela) are unknown in Africa and, except for a comparatively recent penetration into northern South America, they were supposed to be an entirely Holarctic group. But, Milner (1983) has referred to the discovery of urodele fossils in Israel (Lower Cretaceous) and Niger (Cretaceous) as a penetration into Gondwanaland that occurred via a Jurassic European – African land bridge.

The frog family Pipidae is an ancient, archaeobatrachian group that probably evolved in the Jurassic. Fossils are known from the Cretaceous of Israel, Africa, and South America. Its place of origin is unknown but most of the derived forms are found in South America. Living species are confined to Africa and South America. Among the modern neobatrachian frog families, only the Bufonidae, Microhylidae, Ranidae, and Leptodactylidae are found in both Africa and South America (Duellman, 1979).

The fossil record (Paleocene of Brazil) and the rich Neotropical radiation of the Bufonidae indicates that this family originated in South America (Blair, 1972). Although the bufonids are widely distributed, their best development outside South America is in Africa (seven genera and about 50 species). Laurent's (1979) conclusion that the bufonids reached Africa directly from South America rather than by a route through North America and Asia, is reasonable in view of the extensive radiation of the group in Africa.

The history of the family Microhylidae has been a controversial issue among several herpetologists. Continental Africa has three distinct subfamilies with 8 genera, Madagascar has two endemic subfamilies with 11 genera, and a significant radiation has occurred on New Guinea with two subfamilies and 13 genera. The microhylid fauna of South America is quite diverse at the generic level (13 genera) but it, together with the North American forms, is included in one subfamily. Tyler (1979) suggested that the New Guinea fauna developed as the result of an early invasion from the Oriental Region. Laurent (1979) noted that the two most primitive subfamilies occur in Madagascar and southeast Asia respectively.

Savage (1973) felt that microhylid stocks must have once been present on the tropical portions of all the principal southern land masses. However, their virtual absence in Australia and rather weak presence (one subfamily) in the New World does not seem to support this concept. The evidence of a long-term development in Africa (three subfamilies) plus the presence of two relict subfamilies in Madagascar, and the geographic positions of the two most primitive subfamilies, indicates that Africa has probably served as the primary area of evolutionary radiation. As Savage has noted, the microhylids are a relatively primitive family. They were probably more diverse in Africa until the subsequent radiation of the family Ranidae eliminated the microhylids from the tropical lowland habitats.

The family Ranidae seems to clearly be African in its origin and diversification (Savage, 1973). From Africa, several ranoid stocks dispersed to other areas reaching Europe as early as the Eocene and North America by the Miocene. They also invaded the Indo-Australian Archipelago and eventually reached northern Australia and Fiji. The Sooglossidae, an endemic family on the Seychelles Islands, is evidently a relict of the African line that gave rise to the ranoids. There is no direct ranoid relationship to South America. The latter has only a single species that apparently entered from Central America after the establishment of the isthmian link (Duellman, 1979). The family Hyperoliidae originated in Africa and underwent a major radiation (two subfamilies and 18 genera; Savage, 1973). An early stock reached Madagascar but it is found nowhere else. The family Leptodactylidae is represented by a small group of six species in a single genus, all confined to the southern part of Africa; its relationships at the subfamily level (but sometimes placed in a separate family) are with South America and Australia.

As noted (p. 86), the ancient freshwater turtle family Pelomedusidae has a South American – African – Madagascaran range at present but fossils have been found in the Eocene of Asia. Also, it is possible that some older (Triassic and Jurassic) fossils from Germany may be ancestral to this family (Laurent, 1979). There are three modern genera, one shared between Madagascar and South America and the other two confined to Africa. The modern turtle families Testudinidae and Trionychidae occur in Africa but have extensive fossil histories in the northern hemisphere. They are known to have been in Africa since the Miocene and may have entered the continent at that time.

The old lizard family Gekkonidae has already been discussed (p. 86). The lizard family Agamidae is most diverse in the Oriental Region and probably originated there (Briggs, 1984a). There are only three genera in Africa and they evidently represent an invasion from Eurasia (Laurent, 1979). The family Chamaeleontidae is a group of specialized, arboreal lizards probably derived from the agamids. There are six genera and more than 80 species ranging from Africa and Madagascar to southern India and Sri-Lanka. One species is found on some of the Mediterranean islands and ranges north into southern Spain and Portugal (Goin and Goin, 1971). The great majority of the genera and species are confined to Africa or Madagascar and that general area is evidently the primary center for the family.

The family Scincidae (skinks) is a very large group comprising about 80 genera and more than 1000 species (Greer, 1970). All four of the subfamilies are found in Africa and two of them are endemic. Although the group is almost cosmopolitan in distribution, it seems clear that it has undergone most of its evolutionary history in Africa. The family Lacertidae is spread over large areas of the Old World but does not occur in Madagascar or in the Australian region. Lacertids are most abundant in Africa and comparatively rare in the Oriental Region. They may have arisen in Africa.

The worm lizards or amphisbaenians are an old group with fossils known from the early Tertiary of both Europe and North America. One family (Trogonophidae) is restricted to North Africa and Socotra Island in the Indian Ocean. The other family (the Amphisbaenidae) is widespread extending north to southern Europe and to the New World. Altogether, there are about 10 genera and 52 species of amphisbaenians in Africa and 6 genera and 45 species in South America. The generic relationships suggest past exchanges between the two continents.

As was indicated in the section on South America relationships (p. 86), two small snake families, the slender blind snakes (Leptotyphlopidae) and the blind snakes (Typhlopidae), are archaic groups that are widespread in both hemispheres. Both occur in Africa and South America but the relationship is not necessarily a direct one. The family Boidae is an old snake group with an extensive fossil record beginning in the Upper Cretaceous (Underwood, 1976; Parker, 1977); two of the seven subfamilies have complementary distributions: the Boinae are numerous in the Neotropics, absent from Africa, but present again on Madagascar (2 genera, 3 species), skip the Oriental Region, but are found again in Melanesia and Polynesia. The Pythoninae occur from Africa through the Oriental and Australian regions. The subfamily Erycinae includes two genera in western North America and two from northeastern Africa to New Guinea. The Calabar ground python from West Africa is placed in a subfamily of its own (Calabariinae). The distributional patterns of the relicts and the current range of the Pythoninae indicate that most of the family evolution has taken place in the Old World tropics (both in Africa and the Oriental Region).

The Colubridae is a large (more than 2000 species) and relatively advanced family of snakes. The geographical distribution of the most primitive colubrids is predominately in the Neotropical, Nearctic, and Palearctic Regions while the most derived species are concentrated in the Ethiopian and Oriental Regions (Rabb and Marx, 1973). The African and South American stocks seem to be separated at the subfamily level (Duellman, 1979) but may have had some common ancestors. Successions of colubrids probably originated in the Old World tropics and, as they moved out from those areas, eliminated or forced out to the geographical periphery, their more primitive relatives (Rabb and Marx, 1973). It may be noted that there are 15 genera of colubrids native to Madagascar and that all but one are endemic to that island; they are considered to be relatively primitive compared to the African genera.

The family Elapidae includes the cobras and coral snakes. It is most diverse in the Oriental Region and in Australia but extends, in lesser numbers, to Africa (Parker, 1977). The family also extends north into Eurasia and westward to the Nearctic. The general pattern of distribution appears to indicate an evolutionary center in the Oriental Region.

The snakes with the most effective method for injecting venom are the vipers (Viperidae). The group includes the true vipers, pit vipers, and rattlesnakes. The true Vipers (Viperinae) are restricted to the Old World. Two of the most primitive genera of this group are virtually confined to the Ethiopian Region. The majority of the modern genera also occur in Africa but Marx and Rabb (1965) felt that, on the basis of the worldwide pattern, the place of origin appeared to be the Orient or the southeastern Palearctic.

The African tropics support a diverse and evolutionarily advanced bird fauna that is closely related to the bird fauna of the Oriental Region. There are 73 families (6 of them endemic) and about 1556 species (Welty, 1979). Mayr (1976) noted the remarkable lack of relationship between the bird faunas of Africa and South America. Aside from five families of freshwater birds and three families of land birds that are considered to be pan-tropical, there is very little resemblance. There is an interesting, and probably old, migratory relationship between the Ethiopian and Palearctic Regions. In the northern winter, about one-third of all Palearctic bird species, especially the insect-eaters, migrate to Africa south of the Sahara Desert.

The oscine passeriforms or songbirds number about 4000 species and are considered to be the most advanced and successful of all major bird groups. They dominate the avian fauna of the Ethiopian, Oriental, Palearctic and Nearctic Regions. Mayr (1946, 1976) and Darlington (1957) considered that most of the evolutionary development took place in the Old World tropics and that the New World has received the majority of its songbird fauna from the Old World. Also, Darlington suggested that, over the world as a whole, the suboscines (more primitive passeriforms) are being replaced by the oscines.

The mammal fauna of the Ethiopian Region comprises approximately 840 species belonging to 52 families and 14 orders (Bigalke, 1978). Its diversity is quite comparable to that of the Neotropics which includes 810 species, 50 families, and 12 orders (Hershkowitz, 1972). Contemporary African mammals are the products of extensive periods of isolation broken by contacts with Eurasia that permitted interchange during the Cenozoic. Some 15 families and two subfamilies are absolutely or virtually endemic. Ten are shared only with the Oriental Region. Seventeen families also occur in Eurasia (a few also in other regions) and eight are widely distributed.

The Mesozoic record of mammals in Africa is very sparse but the two Triassic forms that are known appear to demonstrate a close faunal tie between Africa and Europe. In the Paleocene and Eocene, it is likely that at least three groups immigrated from the north – one or more primitive condylarthran stocks, prosimian primates, and creodont carnivores (Maglio, 1978). However, as noted below (p.

106), there is a good case, on biogeographic evidence, that the primates originated in Africa. By the early Oligocene, 14 new families appeared, many of them probably endemic in origin. Among them were three primate families, Parapithecidae, Pongidae, and Hylobatidae (under the most recent classification, the latter two would be placed in the family Hominidae).

The most dramatic faunal upheaval occurred in the late Oligocene/early Miocene. Fourteen families recorded earlier are still present but 29 new families and 79 new genera make their appearance. There was evidently a stronger link between Africa and Asia during this time than between Africa and Europe. The Proboscidea began a radiation into several basic groups, archaid suids arrived from Eurasia, and so did canids, viverrids, and felids. By the late Miocene, another 18 families appeared. Artiodactyla began, for the first time, to dominate the African landscape. Nine distinct tribes of Bovidae are recognized, mostly with Asiatic ties (Maglio, 1978).

By Pliocene time, faunal changes at the family level had tapered off with only three new families appearing. Monkeys assignable to the Cercopithecidae became widespread, three new genera of elephants appeared, and 15 new bovid genera. New groups which had recently arrived in Africa included the camel, the modern giraffe, the horse genus *Equus*, and a saber-toothed cat (Maglio, 1978). The early Pleistocene saw three families enter from Europe, the Leporidae, Rhizopodidae, and Cervidae. About 53 new genera appeared, about half being immigrants from the north and half the products of in situ evolution.

The present and fossil distribution patterns of the primates indicate several distinct levels of evolutionary radiation. Immunological data (Cronin and Sarich, 1980) indicate that, of the various stem insectivore groups leading to the primates proper, the flying lemurs (Cynocephalidae) and the tree shrews (Tupaiidae) are the closest. Both families have an apparent relict distribution in southeast Asia but a fossil family, related to the flying lemurs, has been recorded from the late Paleocene and early Eocene of North America (Findley, 1967).

The most primitive living primates belong to the Infraorder Lemuriformes (Szalay and Delson, 1979). Within this group are four families, Lemuridae, Indriidae, Daubentoniidae, and Cheirogaleidae, that are all confined to Madagascar and its vicinity. Only the most advanced family of the Infraorder, the Lorisidae, is found on continental Africa and southeast Asia. It seems clear that, as early primate evolution proceeded in Africa, various groups were able to reach Madagascar, probably by waif dispersal.

The next major step in primate evolution involved the Infraorder Tarsiiformes. The only living remnant of this stage is the genus *Tarsius* (family Tarsiidae), which has a relict distribution in southeast Asia. However, the related family Omomyidae is known from the early Eocene to late Oligocene deposits of North America and Western Europe (Szalay and Delson, 1979). The omomyids may have undergone a waif dispersal to South America in the Eocene (McKenna, 1980) to begin the evolution of the New World monkeys. The latter belong to the Infraorder Platyrrhini.

All of the highest primates are placed in the Infraorder Catarrhini. The Old World monkeys, family Cercopithecidae, are a highly successful primate group. There are 11 living genera, 13 known extinct genera, and well over 100 living species (Szalay and Delson, 1979). The family ranges widely from Africa to southeast Asia and as far north as Japan. The fossil record extends back to the early Miocene. The most recent stage of primate evolution produced the creatures known as the great apes and man. They belong to the family Hominidae. The two more primitive forms are the gibbons (*Hylobates*) and the orangutan (*Pongo*); both are restricted to southeast Asia with *Pongo* being found only on Borneo and Java. The more advanced genus *Pan*, which includes both the chimpanzee and the gorillas, is found on continental Africa.

Fossil genera considered to be ancestral to *Homo* are *Ramapithecus* and *Australopithecus* (Szalay and Delson, 1979). *Ramapithecus* is known from the middle and late Miocene deposits of Turkey, Kenya, India, Pakistan, and China. It may be best considered a sister taxon to *Australopithecus* rather than a direct ancestor. The latter is known from the late Pliocene to early Pleistocene of southern and eastern Africa. The oldest recognized species of *Homo* is generally *H. habilis,* although there is a good deal of controversy about this matter. *H. habilis* is known from the late Pliocene of Africa. *H. erectus,* a more advanced form from the Plio-Pleistocene, is the first member of the genus known to have exited from Africa (to China and southeast Asia).

Modern man is called *Homo sapiens* and a number of fossils originally considered to be separate species, are now placed as subspecies of *H. sapiens* (Szalay and Delson, 1979). Neanderthal man, Heidelberg man, and Rhodesian man are in this category. *H. s. sapiens* refers to Cro-Magnon man and his descendents. All of these forms, if indeed that many are recognizable, lived during the Pleistocene. *H. s. sapiens* survived the ice ages to spread over the world. The earliest definite fossils of *H. s. sapiens* were found in Africa below the Sahara.

The Old World monkeys may have originated in Africa, at least that is where the earliest fossils are from, but soon became more widespread. The distributional pattern of the great apes is indicative of an African genesis. The earliest fossil material is from there and the more advanced forms (gorillas and chimpanzees) are found there today while the more primitive forms (gibbons, orangutan) are relegated to the geographical periphery. It may also be significant that *Australopithecus* and the earliest fossil remains of the genus *Homo* are primarily from Africa. Although the fossil evidence is by no means conclusive it, taken together with the phylogenetic and distributional patterns of the great apes, does favor Africa as the primary area of evolutionary radiation for the primates (including man).

The freshwater fauna of Africa, especially the southern part, exhibits some interesting relationships to both South America and to the Australian – New Zealand area. The mussels and snails have been discussed (p. 89). The brachyuran family Pomatomidae has its evident center of origin in southeast Asia but there is a significant secondary radiation in Africa (Bishop, 1967). In the parasitic nematode family Camallanidae (Stromberg and Crites, 1974), the African fauna appears to have been derived from southeast Asia and the South American fauna, which is comparatively primitive and depauperate, was probably derived from Africa in the Mesozoic.

The relationships of other southern African freshwater groups has been reviewed by Harrison (1978). He considered the fauna to be divisible into two major groups, the South Temperate Gondwanian and the Pan-Ethiopian. Examples of the former are some primitive southern stoneflies (Plecoptera) of the family Notonemouridae surviving in southern Africa and Madagascar and several genera of the mayfly (Ephemeroptera) family Leptophlebiidae. Besch (1968) called attention to the relationship of the water mite (Hydracnellae) genus *Australiobates* which is found in South Africa, southern South America, and Australia. The Pan-Ethiopian fauna would include groups such as the freshwater ostracods that Grekoff and Krommelbein (1967) discovered in the Upper Jurassic strata of northeastern Brazil and West Africa.

Edmunds (1972), in reviewing the biogeography and evolution of the mayflies, noted that the distribution of many groups indicated a tropical connection or near-connection of Africa and South America and that there were numerous invasions of the Oriental Region (and to a lesser extent of the Palearctic and Australian Regions) by Ethiopian mayflies. Other aquatic groups show similar distributions: the caddisfly (Trichoptera) genus *Leptonema* is known from Africa and the Neotropics; the freshwater snail genus *Biomphalaria* (Pulmonata: Planorbiidae) is known from the West Indies, South America, Africa, and Madagascar; and the Torridincolidae (Coleoptera) are found in Brazil, southern Africa and Madagascar (Harrison, 1978).

The distribution of the freshwater ostariophysan fishes was discussed earlier (p. 91). It was suggested that the early characoids, followed closely by the first siluriform fishes, originated in the Oriental Region in the Upper Jurassic. From tropical Asia, these groups probably spread to Europe via an occasional trans-Turgai connection. Once in Europe, they could have migrated around the west end of the still incomplete Tethys Sea and entered Africa, and thence South America. The cyprinoid fishes, the most advanced of the ostariophysan groups, evidently entered Africa too late to reach South America.

The cyprinoid fishes comprise a huge group of six families, 256 genera, and about 2422 species (Nelson, 1984). The greatest diversity is in southeast Asia where all six families occur. Two families, the Cyprinidae and the Cobitididae, are found in African freshwaters. Fossils of cyprinid fish from Kenya have been dated from the early Miocene and it has been suggested that these fishes entered Africa about 18 – 16 Ma (Van Couvering, 1977). The Oriental Region has possibly also been the place of origin of other freshwater fish families (Anabantidae, Channidae, Mastacembelidae) that eventually made their way to Africa (Briggs, 1979).

As far as other groups of primary freshwater fishes are concerned, Africa has a far more diverse fauna than South America. The families Polypteridae, Denticipitidae, Pantodontidae, Phractolaemidae, Kneriidae, Mormyridae, and Gymnarchidae are all endemic to Africa. These are very old, primitive families that probably date back to the early Mesozoic (Briggs, 1979). Only one, the Pantodontidae, has possible relatives from the marine environment. South America has few groups of primary freshwater fishes other than ostariophysans.

The secondary freshwater fish order Cyprinodontiformes evidently arose in the New World (p. 92). A primitive species could have reached Africa in the early Cretaceous when that continent was still close to South America. The greatest species and generic diversity in the Old World is found in Africa and, from there, representatives were probably able to migrate eastward to Madagascar and the Seychelles and northward to Eurasia.

As noted in the South America discussion (p. 94), some primitive, terrestrial insect groups appear to demonstrate southern continent relationships. Southern Africa is involved in a few such distributions but not as strongly as South America. Both continents have large termite (Isoptera) faunas but the major center of evolution has probably been tropical Africa (Emerson, 1955). The Ethiopian termite fauna is most closely related to that of the Oriental Region (Bouillion, 1970). Of 89 Ethiopian genera, 60 are endemic. Within the large family Termitidae, one subfamily is apparently of Oriental origin, two probably arose in Africa, and another developed in the Neotropics.

Brown (1973) noted that the ant faunas of sub-Saharan Africa and the Neotropical Region, including those of the rain forests, are very different from one another at the species group and generic levels. The two regions share only 29 genera, all widespread in the tropics; most of them also extend to the northern hemisphere either now or as Tertiary fossils. From at least mid-Tertiary times, evolution of world-dominating new taxa has proceeded mainly from combined tropical Africa – southern Asia. Genera or generic groups well represented in the Neotropical and Indo-Australian regions, and absent or very rare in Africa, are the peripheral relicts of older taxal waves that are now being replaced from the Old World tropics.

The origin and dispersal of the land plants, particularly the very old groups, provide vital information about continental relationships. Schuster (1969, 1972) has discussed the distribution of the liverworts (Hepaticae), an ancient group in which certain modern genera have close counterparts in the Carboniferous. Since the general sub-antarctic region contains an unusually high preponderance of primitive genera of Hepaticae (about 50%), Schuster concluded that many of the major groups originated in Gondwanaland and that the present relict patterns are the result of continental drift and accompanying climatic changes.

The distribution maps of the primitive Hepaticae that have been published by Schuster (1969, 1972) do not provide support for the idea that these plants had a widespread Gondwanian origin. The maps indicate a high concentration of relicts in the Antipodes area (New Zealand, Tasmania, southeastern Australia, New Caledonia) and a significant but lesser number in southern South America. Such relicts are almost entirely missing from southern Africa and India. So, while an Antipodes – South American relationship is demonstrated, the other supposed Gondwanaland continents are not involved.

Delevoryas (1973) pointed out that, during the Permian and Triassic periods, there were many primitive vascular plants that had very broad distributions and that a fair number of them were restricted to the southern hemisphere, thus being considered Gondwanian. Several genera such as *Neocalamites, Equisetites,* and *Cladophlebis* have been found in Triassic rocks of both Africa and South America. This impression of floral continuity appeared to continue into the Jurassic, although the fossil record at that time became relatively poor. According to Delevoryas, it was not until the Cretaceous that significant differences developed among the floras of the southern hemisphere.

The distribution of conifer and taxad genera, as analyzed by Florin (1963), offers some additional data. One of the oldest of the living families (late Triassic) is the

Araucariaceae. The recent species of *Araucaria* have an interesting disjunct distribution in the southern hemisphere (Antipodes – South America) but the fossil material indicates a much wider distribution in the Mesozoic and early Tertiary. It once occurred in South Africa, Antarctica, and Kerguelen Island but was also widespread in the northern hemisphere ranging from North America to Greenland, Europe, Russia, and India.

The Podocarpaceae, with seven genera and about 150 species, is the most important among the conifer families represented today in the southern hemisphere. Although the genus *Podocarpus* does occur in Africa and Madagascar, the other genera do not. In the family Cupressaceae, about half the genera occur in the southern hemisphere. Of the southern genera, seven are found in the Antipodes area, three in southern South America, and one in southern Africa. Florin (1963) presented a table of the geographical location of the relict conifer and taxad genera (and sections of genera) of the world. None are located exclusively in southern Africa.

The distribution of many of the most primitive dicot (Angiospermae) families has been analyzed and Schuster (1972, 1976) has summarized much of this information. Ten of the 19 families considered by Schuster are restricted to the southern hemisphere but none of these supposedly Gondwanic groups occurs on continental Africa. One, also shared with South America, occurs on Madagascar. The single beech genus of the southern hemisphere is *Nothofagus* (Humphries, 1981). By late Cretaceous times, *Nothofagus* had apparently spread over an area including New Zealand, Antarctica, Australia, Tasmania, and southern South America. It apparently has never lived in Africa.

Although the angiosperm flora of Africa is quite varied, it is not nearly as distinctive as that of South America. In tropical Africa, 17 families are endemic. If temperate southern Africa and Madagascar are included, eight additional endemic families may be added. This may be compared to 47 endemic families in the Neotropical Region. Although about 60 families have pantropical distributions, only about 12 are shared exclusively between Africa and South America (Good, 1974). There are a few families that are well represented on all the southern continents. One of these is the Proteaceae. Despite its generally broad distribution, it may be noted that about two-thirds of the species are Australian and about one-quarter African.

In discussing the relationship of the angiosperm floras of Africa and South America, Smith (1973) presented a generalized model of angiosperm evolution showing a southeastern Asian center of origin and various migration routes leading to other parts of the world. The one major route to Africa was along the northern border of the Indian Ocean and across the Arabian Peninsula. Two major routes led to South America: one via northeast Asia, across the Bering Land Bridge, and south through Central America; the other led down through Australia – New Zealand and across Antarctica. In the vast majority of the ranalean families studied by Smith, he found no relationship between African and South American elements that would indicate a direct interchange of genetic components.

Thorne (1973) also called attention to the lack of relationship between the seed plants of Africa and South America. He concluded that the botanical evidence argued against continental drift as a significant factor in explaining the distribution of angiosperms between Africa and South America and that the separation of the two continents must have attained its present state, or nearly so, before the development of the present seed floras of the world. Niklas (1981) expressed the opinion that continental disjunction had occurred well before the development of present seed-plant floras. Richards (1973) recognized the relatively depauperate condition of the African angiosperm flora compared to that of South America and was inclined to attribute the difference to Africa's more extreme climatic history.

The opinions about the nature of the African flora and its relationship to that of South America that were expressed by Smith (1973), Thorne (1973) and Niklas (1981) were paralleled by Brenan (1978). The latter emphasized the poverty in or complete absence from tropical Africa of many plant groups that might be expected to be well represented. Among them are many large families that have numerous species in America and Asia. Several reasons for this floral poverty were given including the effects of drought, orogenic increases in altitude, development of the Benguela Current, and major fluctuations in the Quaternary climate, as noted by Raven and Axelrod (1974).

In considering the African flora in general, attention should be paid to the Cape Floristic Region (Goldblatt, 1978) for within this relatively small area are some 8550 species in 957 genera. This rich and diverse flora may be contrasted to the relatively depauperate tropical flora of Africa. The Cape flora is believed to have evolved gradually since the early to mid-Tertiary, partly from an ancient southern African flora and partly from the tropical African forest flora. In the Cape region, 6 families, 198 genera, and 6252 species are considered to be endemic. There is an unusually high proportion of petaloid monocots, many succulents, and large numbers of sclerophyllous to microphyllous shrubs.

AFRICA SUMMARY

Although Africa probably lost its tenuous connection to South America in the late Jurassic, it continued its close relationship with Eurasia and, at various times during the Tertiary, major biotic exchanges took place across a narrowed or land-bridged Tethys Sea. The southern tip of Africa supports a few groups of primitive animals and plants that demonstrate southern continent relationships. However, these relationships are not nearly as strong as those shown by the southern part of South America. To the east, Madagascar has served as the recipient of many African forms that were dominant at various times from the mid-Mesozoic through the Cenozoic. Thus, Madagascar serves to some extent as a living museum where the biotic history of Africa may be reviewed.

The geographical relationships of the modern African amphibian fauna are still obscure. Due to the superior family level diversity of the caecilians and frogs in South America, it is tempting to speculate that Africa received its early amphibian fauna from that source. Fossils of the primitive leiopelmatid and ascaphid frogs have been described from the Jurassic of South America. However, the fossil of a proanuran or "prefrog" has been found in the Lower Triassic deposits of Madagascar (Savage, 1973). Since most hypotheses of amphibian relationships treat the Anura and Urodela as sister groups (Milner, 1983), the existence of the Madagascar fossil anuran indicates that a urodele stem-group had also evolved by that time. Judging from contemporary and fossil distribution patterns as they are now known, it appears likely that the frogs underwent their early development in the southern hemisphere, with the emphasis on South America, while the salamanders developed in the northern hemisphere with the emphasis on North America.

The ancient freshwater turtle family Pelomedusidae has a South American – African – Madagascar range but this does not indicate a strictly southern hemisphere history. This and the more modern turtle families Testudinidae and Trionychidae have extensive fossil histories in the northern hemisphere and are unknown in Africa before the Miocene. They may have entered from Eurasia at that time.

The African lizard fauna is of mixed origins. The old family Gekkonidae may have originated in southeast Asia and reached Africa sometime in the Upper Jurassic. Subsequently, it became almost cosmopolitan in the warmer parts of the world. Another primitive lizard family, the Iguanidae, probably originated in South America and, at one time, almost certainly existed in Africa since there are two genera isolated on Madagascar. The modern family Agamidae probably originated in the Oriental Region. It is not very diverse in Africa. However, there arose in Africa, probably from an agamid ancestor, the Chamaeleontidae, a group of highly specialized, arboreal lizards.

The lizard family Scincidae is very large and widely distributed and it seems clear that it has undergone most of its evolutionary history in Africa. In a like manner, the family Lacertidae, which is found over large areas of the Old World, probably originated in Africa. The worm lizards or amphisbaenians are very diverse in both South America and Africa and there have evidently been past exchanges between the two continents.

The primitive snake family Boidae contains two large subfamilies that have complimentary distributions. The Boinae are numerous in the Neotropics, skip Africa, are present again on Madagascar, bypass the Oriental Region, but are found again in certain isolated localities in Polynesia and Melanesia. The Pythoninae extend from Africa to the Oriental Region and out to Australia. The distribution and relationship of the other subfamilies indicates major evolutionary activity in both the Ethiopian and Oriental Regions. The relationships of the large family Colubridae also indicate a major evolutionary activity in the Old World tropics with dispersals to other parts of the world. In terms of South American affinities, both families evidently reached that continent from Africa.

The snake family Elapidae is most diverse in the Oriental and Australian Regions and extends, in lesser numbers, to Africa. Its center of origin is probably the Oriental Region and it extends far to the north in Eurasia. It probably reached the New World via the Bering Land Bridge rather than migrating via South America. The same general pattern is evident in the most advanced snake family, the Viperidae. There are three South American genera of Crotalinae, a viperid subfamily, but it is clear that they arrived there via North America.

In contrast to the amphibians, lizards, and snakes, the African bird fauna shows very little relationship to that of South America. Aside from a few families that have pan-tropical distributions, the great majority are completely different. In general, the bird fauna of Africa, which contains large numbers of the advanced oscine passeriforms, is very close to that of the Oriental region.

The Ethiopian Region contains a rich mammalian fauna that is comparable in diversity, but unrelated, to that of South America. Beginning in the Mesozoic, Africa apparently was invaded by some very early mammals including condylarths, insectivores, and creodont carnivores. By the early Oligocene, 14 new families appeared, many of them probably of African origin. In the late Oligocene/early Miocene, a major disturbance took place when 29 new families made their appearance. This reflected a strong migratory movement from Europe. By the late Miocene, another 18 families appeared. The Pliocene and Pleistocene saw new groups arrive in Africa but these times were also important for the export of African mammals to Eurasia.

The presence of the most primitive living primates (four families of lemurs) on Madagascar plus a fifth family (Lorisidae), the most advanced of the lemuriform group, on continental Africa, gives a strong indication that the early steps in primate evolution took place in Africa. Although the Tarsiidae, representing the next step in primate evolution, have a probable relict distribution in the East Indies, a related family, the Omomyidae had a widespread early Tertiary distribution. The latter may have been ancestral to both the New World and Old World monkeys.

The Old World monkeys (Cercopithecidae) probably evolved in Africa but soon became a widespread, successful group. The great apes and man, which belong to the family Hominidae, represent the most recent stage of primate evolution. The two most primitive genera, the gibbons (*Hylobates*) and the orangutan (*Pongo*), are both restricted to southeast Asia but the more advanced chimpanzee and gorillas (*Pan*) inhabit continental Africa.

The oldest recognized species of *Homo, H. habilis,* is known from the late Pliocene of Africa. *H. erectus,* a more advanced species from the Plio-Pleistocene, is the first member of the genus known to have exited from Africa. The earliest definite fossils of *H. s. sapiens,* modern man, were found in Africa below the Sahara. Although the fossil evidence is not conclusive it, taken together with the phylogenetic and distributional patterns of the great apes, does favor Africa as the primary area of evolutionary radiation for the primates (including man).

In the older groups of animals that inhabit the freshwater streams and lakes, one can find some interesting ties to both South America and the Australian – New Zealand area. An African mussel family is related to another family found in South America; a snail family with advanced genera in southeast Asia has its three most primitive genera in South America, South Africa, and Australia, respectively; and a brachyuran family with its center of origin in southeast Asia has a significant secondary radiation in Africa. A parasitic (on freshwater fishes) nematode family apparently originated in the Oriental Region in the Mesozoic then migrated with its hosts to Africa and finally to South America.

Additional early ties to Australia may be found among the stoneflies (Notonemouridae), several mayfly (Leptophlebiidae) genera, and the water mite genus *Australiobates.* Other connections to the Neotropics are present in several groups of mayflies, the caddisfly genus *Leptonema,* the snail genus *Biomphalaria,* and the beetle family Torridincolidae.

The ostarophysan fishes, a primary freshwater group with a long history in that environment, are of great biogeographic importance. The Cyprinoids (minnows), the most advanced ostariophysan group, are found in Africa (two families) but not South America. The cyprinoid center of diversity and evolutionary radiation is located in the Oriental Region. From there, they entered Africa probably in the early Miocene.

Africa has, at the higher taxonomic levels, a far more diverse primary freshwater fish fauna than South America. Such families as the Polypteridae, Denticepitidae, Pantodontidae, Phractolaemidae, Kneriidae, Mormyridae, and Gymnarchidae are all old families probably dating back to the Mesozoic; yet, none of them is found in South America (or elsewhere). The secondary freshwater fish order Cyprinodontiformes evidently arose in the New World. Many of the species are euryhaline so that a migration to Africa from South America in the early Cretaceous would have been possible across a narrow saltwater barrier.

In contrast to the various groups of freshwater invertebrates and fishes, the terrestrial invertebrates of Africa do not demonstrate much relationship to those of South America. The land snails do not show any particular affinities. Among the termites, Africa has served as the main evolutionary center and the fauna in general is closest to that of the Oriental Region. One primitive termite genus is represented by one species each in southern Africa, southern South America, and Australia – Tasmania. The most advanced groups of ants are found in Africa and the Oriental Region while those of the Neotropics and Indo-Australia are considered to be more primitive.

The African flora, both fossil and recent, does not demonstrate much in the way of Gondwanaland relationships. Very few of the most primitive liverworts (Hepaticae) are present. A few vascular plant fossils have been taken from Triassic deposits of both Africa and South America giving some indication that the floras at that time may have been quite similar. Relict conifer and taxad genera that are numerous in Australasia and South America are notably absent from Africa.

Among the angiosperms, Africa possesses none of the families that are considered to be the most primitive. Even, *Nothofagus*, the southern beech, which is otherwise widespread in the southern hemisphere, does not occur in Africa. In regard to the higher angiosperms, the African flora has close ties to that of southeast Asia but is very different from that of South America. About 60 families have pan-tropical distributions but only about 12 are shared exclusively between Africa and South America. The evolutionary center for angiosperms was probably southeast Asia. From there, Africa probably received most of its flora by means of a migration route along the northern border of the Indian Ocean and across the Arabian Peninsula.

The flora of tropical Africa, compared to that of the Neotropical and Oriental Regions, is depauperate. The reasons given include the effects of drought, orogenic increases in altitude, development of the Benguela Current, and major fluctuations in the Quaternary climate. In contrast, the flora of the Cape Region is unusually diverse consisting of some 8550 species in 957 genera. This flora is believed to have evolved gradually, partly from the tropical African forests and partly from an ancient southern Africa flora. In the Cape region, six families, 198 genera, and 6252 species are considered to be endemic, giving the region a remarkable distinctiveness.

Chapter 11

MADAGASCAR

> . . . archaic forms, which among modern groups resemble most closely the putative ancestors of major taxonomic categories, are most likely to be preserved under conditions that are least favorable for the origin of new, more efficient genotypes that would compete successfully with them and drive them out. This . . . emphasizes the necessity for distinguishing clearly between biotic communities that function as evolutionary "laboratories" or "cradles", in which new adaptive complexes arise, and "museums" in which archaic forms are preserved.
>
> G.L. Stebbins, in: *Ecological Distribution of Centers of Major Adaptive Radiation in Angiosperms,* 1972

Although some continental reconstructions for the Mesozoic show Madagascar sandwiched between India and east Africa, and forming a land bridge between the two, until the late Cretaceous (Colbert, 1980; Hallam, 1981), the biological relationships of Madagascar reflect an isolation that probably extended back to much earlier times. Upper Permian fossil material from Madagascar indicates the presence of procolophonoids, lepidosaurs, and phytosaurs. From the Middle Jurassic to the Upper Cretaceous, there were various dinosaurs such as the titanosaurs, megalosaurs, and stegosaurs (Blanc, 1972). In general, the geographic affinities of these forms were widespread involving Africa, Europe, Asia, and North America. Cox (1974) suggested that the Cretaceous dinosaur fauna was descended from a common Jurassic stock and that the new Cretaceous families developed in the north were unable to reach the southern hemisphere. However, Sues and Taquet (1979) reported an advanced pachycephalosaurid dinosaur with northern relationships from the Upper Cretaceous of Madagascar.

The earliest known stem-anuran, *Triadobatrachus* has been taken from Lower Triassic deposits on Madagascar. It shows evidence of ancestry among Paleozoic amphibians (Estes and Reig, 1973). The modern herpetofauna of Madagascar shows some interesting relationships. Africa has been suggested (p. 101) as the most likely area for the origin of the frog family Microhylidae. Three subfamilies are found in Africa and two others on Madagascar. One of the Madagascar subfamilies is considered to be relatively primitive while the other is more advanced. Africa has a total of eight genera while Madagascar has eleven. Savage (1973) felt that the microhylids were probably more diverse in Africa at one time but that their numbers were reduced by the subsequent radiation of the family Ranidae. It is curious that the caecilians (Caeciliidae) and the frog family Sooglossidae (an offshoot of the line that led to the Ranidae) should occur on the Seychelles Islands but not on Madagascar itself.

The ranid treefrog subfamily Rhacophorinae is represented on Madagascar by a primitive genus (*Boophis*) with about 50 species (Savage, 1973). There is only one genus with 12 species in Africa. Savage noted that an elimination of the more generalized rhacophorines in Africa coincided with the rise of the hyperoliid tree frogs. Fairly early in the Cenozoic, another stock of primitive ranids reached Madagascar where they differentiated into the endemic subfamily Mantellinae (4 genera, 50 species). The family Hyperoliidae evidently originated in Africa to become a major group (two subfamilies, 18 genera, and about 300 species). A hyperoliid derivative stock (2 genera, 10 species) apparently arrived on Madagascar in the early Tertiary while two more recent immigrations involved modern Africa genera (two genera and possibly 12 species).

It has been noted (p. 102) that the ancient freshwater turtle family Pelomedusidae has a South American – African – Madagascar range at present but that fossils have been found in the Cretaceous of Europe and North America and in the Eocene of Asia. One of the three modern genera (*Podocnemis*) is found in both Madagascar and South America but not in Africa. In the modern turtle family Testudinidae, there is a group of giant terrestrial tortoises (Testudininae) with a disjunct distribution occurring on Madagascar – Seychelles – Aldabra and then also on the Galapagos Islands. A fossil species of this group is known from the Eocene of Egypt (Blanc, 1972). However, the giant species are related to a smaller *Testudo* found in southern Europe, Asia, and Africa so that the disjunction probably represents the remnants of a formerly continuous range.

The old lizard family Gekkonidae is a large group of about 75 genera and hundreds of species. It has a circum-tropical distribution and the individual species are adept at waif dispersal. On Madagascar there are 14 genera and about 70 species (Blanc, 1972). The lizard family Iguanidae is highly developed in the New World, does not exist in Africa, but is represented by two genera on Madagascar. At one time, the family must have existed in Africa. It has been suggested (p. 86) that their extinction in Africa may have been due to competition from the more advanced agamid and chamaeleontid lizards. The family Agamidae is not found on Madagascar but the Chameleontidae is represented by a fauna of 4 genera and about 35 species. This could mean that the competition provided to the iguanids by the agamids was more important than that provided by the chamaeleontids.

The large, cosmopolitan lizard family Scincidae (skinks) has probably undergone most of its evolutionary development in Africa; all four of the subfamilies are found there and two of them are endemic (Greer, 1970). There are 10 genera and 51 species on Madagascar (Blanc, 1972). The Cordylidae is a small family of lizards, probably derived from the skinks, that is confined to Africa and Madagascar. There are 2 genera and 13 species in the latter area. The advanced family Lacertidae probably arose in Africa and is spread over large parts of the Old World but is not found on Madagascar. The worm lizards (Amphisbaenidae) are also widespread but do not occur on Madagascar.

Among the families of snakes, the blind snakes (Typhlopidae) are an archaic group that is widely distributed in both the New and Old World. Madagascar has one genus and nine species. The family Boidae is also an old snake group, two of the subfamilies have complementary distributions: the Boinae are numerous in the Neotropics, absent from Africa, present again on Madagascar (2 genera, 3 species), skip the Oriental Region, and are found again in isolated localities in Melanesia and Polynesia (Parker, 1977). The Pythoninae occur from Africa through the Oriental and Australian Regions but not on Madagascar. It appears that the pythons may have eliminated the boines from Africa and elsewhere leaving the latter group with a fragmented distribution. The only other snake family found on Madagascar is the Colubridae, a large cosmopolitan group (Rabb and Marx, 1973); there are 15 genera native to Madagascar and all but one are endemic; compared to the African genera, they are relatively primitive. It may be noted that the advanced families Elopidae (cobras and coral snakes) and Viperidae (vipers) are not found on Madagascar.

The bird fauna of Madagascar consists of 238 species in 164 genera (Dorst, 1972). In comparison to other large islands, such as New Guinea with about 650 species, the diversity of Madagascar is quite low. However, the fauna is interesting in its peculiarities and Dorst considered it to be comprised of elements of an archaic fauna formerly found in many other parts of the world. Such birds can still exist on Madagascar because of the island's isolation and the absence of more advanced competitors. Many characteristic African families are absent; among these are the woodpeckers (Picidae), barbets (Capitonidae), honey-guides (Indicatoridae), trogons (Trogoniformes), hornbills (Bucerotidae), shoebills (Balaenicipitidae), secretary-birds (Sagittariidae), turacos (Musophagidae), wood hoopoes (Phoeniculidae), mousebirds (Coliidae), oxpeckers (Buphagidae), picathartes (Picarthartidae), and promerops (Promeropidae).

The extensive isolation of the "Great Island" has produced many endemic genera and some families. The ancient ratites were at one time represented by an endemic family (Aepyornithidae) with 12 species. Other peculiar families are the ralliform family Mesitornithidae, the Philepittidae, and the shrike family Vangidae. A total of 46 bird genera are restricted to the island and some of them are so ancient that it is difficult to associate them with other systematic entities (Dorst, 1972). In general, it may be said that the avian faunal affinities belong with Africa, but there are about a dozen species that belong to typical Asian genera.

The mammalian fauna of Madagascar is depauperate, generally archaic, and devoid of large herbivorous or carnivorous forms. A temporary exception was a pigmy *Hippopotamus* which apparently reached the island in the Pleistocene and then became extinct in recent times (Mahe, 1972). The shrews (Soricidae) are represented by two endemic and one introduced species. However, the tenrecs (Tenrecidae) are numerous (9 genera, 27 species) and show evidence of considerable evolutionary radiation on Madagascar (Heim de Balsac, 1972). Elsewhere, the family occurs only in tropical West Africa. Fossils are known from the early Miocene of Kenya and the group may have reached Madagascar about that time. There are seven genera and ten species of rodents belonging to the family Cricetidae. All of the genera are endemic and are placed in a distinct subfamily (Nesomyinae). It is not clear whether these genera are indeed closely related or whether each represents a separate family (Petter, F., 1972).

The Carnivora of Madagascar all belong to one family, the Viverridae. These animals are commonly known as civits, genets, and mongooses. There are seven native genera and eight species (Albignac, 1972). The family is widespread in the Old World. The most primitive living primates belong to the Infraorder Lemuriformes (Szalay and Delson, 1979). Four of the five families, the most archaic, are confined to Madagascar. They are the Lemuridae, Indriidae, Daubentoniidae, and the Cheirogaliidae. Only the most advanced family of the Infraorder, the Lorisidae, is found on continental Africa. Fossil lemurs, several of them quite close to the present Madagascar forms, were common and widespread during the Eocene (Petter, J.J., 1972). In the rest of the world, the lemurs rather suddenly disappeared in the

Oligocene leaving only the four families on Madagascar and the lorisids in the Old World tropics. It has been suggested (p. 105), that as early primate evolution proceeded in Africa, various groups were able to reach Madagascar, probably by waif dispersal.

The freshwater fauna of Madagascar is of considerable interest. A number of primitive aquatic insect groups possibly demonstrate old southern continent relationships (Paulain, 1972): the blepharocerid (Diptera) subfamily Edwardsininiae is elsewhere found only in South America and Australia, stoneflies (Plecoptera) of the family Notonemouridae have South African and New Zealand relationships, and two genera of dragonflies (Odonata) have African relatives. The caddisflies (Trichoptera) are represented by an endemic family and several odd and primitive genera (Ross, 1967). The archaic mayfly (Ephemeroptera) family Leptophlebiidae is represented by genera in South America, Australia, New Zealand, Africa, and Madagascar (Pescador and Peters, 1980). The Torridincolidae (Coleoptera) have been found in Brazil, southern Africa, and Madagascar, and the freshwater snail genus *Biomphalaria* (Pulmonata: Planorbiidae) is known from the West Indies, South America, Africa, and Madagascar (Harrison, 1978). A parastacid crayfish genus (*Astracoides*) is related to genera in other parts of the southern hemisphere.

The rich, primary freshwater fish fauna of Africa, which contains a number of old, endemic families, is entirely absent from Madagascar. It has been suggested that the ostariophysan groups Characoidea (characins) and Siluriformes (catfishes) entered Africa from Europe in the late Jurassic (Briggs, 1979). Some of the endemic families are morphologically very primitive and may be as old, or older than, the two ostariophysan groups. But these developments came too late to involve Madagascar. The freshwater fish fauna of Madagascar consists of two secondary freshwater families (groups that prefer freshwater but can tolerate high salinities), the Cichlidae and the Cyprinodontidae, and a large group of marine families that are represented by euryhaline species. The relationships of the cichlids apparently lie with both Africa and India (Kiener and Richard-Vindard, 1972). The cyprinodontids evidently arose in the New World, invaded Africa, and subsequently reached Madagascar (p. 93).

There is some information about the affinities of the terrestrial invertebrate fauna of Madagascar. The molluscs have been discussed by Fischer-Piette and Blanc (1972). Of a total of 38 genera, five were considered to be strictly endemic but the majority of the rest had Oriental or Oriental plus African relationships. Three of the endemic genera had their closest relatives in Sri-Lanka. Only two genera were shared with Africa alone. In his general discussion of the insect fauna, Paulain (1972) noted that southern continent (austral) elements appear sporadically in the most diverse groups but, with the exception of the Scarabaeidae (Coleoptera), they do not appear to constitute an important part of the present Madagascar fauna. As stated (p. 94), a primitive beetle tribe, the Ctenostomini, and other beetle groups are restricted to Madagascar and the Neotropical region (Reichardt, 1979). The same pattern is found in some mayfly lineages (Edmunds, 1982).

In his review of the arachnid fauna of Madagascar, Legendre (1972) emphasized a relationship to Africa but also stated that the old spider family Archaeidae provided a fine example of austral distribution. The various genera may be found on

Australia, New Zealand, Africa, and Madagascar. However, as was the case with some of the beetle groups, the Baltic Amber has produced related specimens so the Archaeidae may have had a tropical or northern origin. Legendre concluded that although African relationships were predominant, some Indo-Pacific influence could be detected.

Sixty-one species of termites (Isoptera) have been recorded from Madagascar (Paulain, 1970). They belong to the three main families, Kalotermitidae, Rhinotermitidae, and Termitidae. The most primitive groups are not found here but neither are the most advanced. Paulain has noted that the termites, as is the case for most other insect orders on Madagascar, seem to have been selected on a haphazard basis; some genera have established themselves in certain habitats but other closely related genera are missing. The percentage of endemic species is very high reaching 100% in the 33 species of Termitidae, 90% in the 23 species of Kalotermitidae, and 40% in the 5 species of Rhinotermitidae. The Madagascar termite fauna is apparently entirely African in origin.

Leroy (1978), in his analysis of the composition and origin of the Madagascar vascular flora, adopted the hypothesis that it was a part of the large Cretaceous Gondwanian flora at the time when the angiosperms originated and that it retained many of its original characteristics while the flora of tropical Africa became greatly impoverished due to far-reaching climatic changes. Due to the presence of the genus *Bubbia* of the primitive angiosperm family Winteraceae, Leroy felt that the Madagascar region was a part of the cradle of the primitive angiosperms. However, a look at the general distribution of the Winteraceae (Smith, 1972) shows that there are seven genera, one occurring in South to Central America and the other six primarily situated in the Australian – New Zealand area. It is one of the latter six (*Bubbia* with 30 species in New Caledonia, New Guinea, and Australia) that also occurs as a single species on Madagascar.

Other families, such as the Myristicaceae and the Canellaceae, are also considered by Leroy (1978) to have Gondwanaland origins. But the former family also occurs in tropical America, Africa, and Asia and may be best considered pan-tropical. The latter occurs also in Africa and America but not in the Australian – New Zealand area. The family Monimiaceae occurs primarily in Madagascar and adjacent islands but there is a monotypic genus living in Chile (Lorence, 1985). Fossils are known from the late Cretaceous of Europe, Africa, and Argentina. Thus, we have another example of a disjunct Madagascar – South American distribution that was probably caused by extinction in Africa (and elsewhere).

Although other, more advanced, angiosperm families have their primary distributions in the southern hemisphere, they are usually considered to be too young to trace their histories back to the pre-drift assemblage of continents. In regard to the early angiosperms, it must be said that the evidence for their development in Madagascar or Africa is remarkably sparse. The concentration of primitive families in the general East Asian to Indo-Australian area is much greater. Even the southern part of South America has a much larger representation of primitive angiosperms in its flora than does Madagascar.

Among the gymnosperms, the Podocarpaceae is an important and old family. One of the seven genera, *Podocarpus* with six endemic species, is present on

Madagascar. However, its presense is not necessarily indicative of Gondwana relationships. Although the family is primarily found in the southern hemisphere, it is a tropical as well as a temperate group. Species of *Podocarpus* are also found in tropical South America, Africa, and East Asia. In the New World they extend north through Mexico and the West Indies and in the Old World they go north to Honshu Island in Japan (Florin, 1963). There are Triassic fossils of another conifer belonging to the genus *Voltziopsis* which also has been reported from continental Africa.

Although Madagascar does not have a good representation of the very primitive angiosperms and gymnosperms, its flora is nevertheless very rich and highly endemic. Perhaps because most plant species are better adapted than most animal species for long distance dispersal, the island flora contains a fine representation of various plant groups that were once common in Africa. So, by studying the Madagascar flora the botanist can become better acquainted with the history of floral succession in Africa. Madagascar together with the Mascarene Islands has about 200 families (12 endemic), about 1150 genera, and some 6000 species (Good, 1974). Another estimate by Koechlin (1972) gave about 10,000 species.

More than 350 of the genera are endemic to Madagascar itself and these contain about 1000 species (Good, 1974). About 35 other genera are endemic to Madagascar *and* to one or more of the island groups in the region. The general proportion of endemic species is about 90%. The largest families in terms of numbers of species are the Orchidaceae (685 species), Compositae (388), Euphorbiaceae (318), Cyperaceae (316), and Gramineae (300). The external floral relationships lie primarily with tropical Africa but there is also a strong pan-tropical element of about 42% of the species. An eastern element of affinity to the floras of eastern Asia and the tropical Pacific amounts to about 7%.

Some of the major plant groups that are still very diverse on Madagascar have become relatively scarce in Africa. The palms (Order Principes) are a large, circumtropical, monocot group consisting of more than 200 genera and over 3000 species. They are very well established in the New World, from which they may have originated, and also in southeast Asia to Australia (Moore, 1973). But, they are very scarce in Africa which has only three genera and a few species. In contrast, Madagascar has 18 genera and over 100 species. A similar distribution may be noted for the bamboos (Bambusaceae), a large tropical group of about 50 genera and 500 species (Good, 1974). About 90% or more of the species are either Asiatic or American. There are only about 6 genera and 14 species in Africa but Madagascar has 9 genera and about 30 species. In both the palms and bamboos, it seems reasonable to suggest that Madagascar has preserved the remnants of an African flora that was, at one time, much richer.

MADAGASCAR SUMMARY

One can summarize the main points about the composition and relationships of the Madagascan biota by looking at it from the standpoint of a traveler who has just left tropical Africa. If our traveler was a zoologist, he would certainly be interested in the strange forms of the native animals but otherwise puzzled by the general depauperate condition of the fauna. He would quickly note large gaps in the faunal continuity giving the strong impression of a disharmonic or unbalanced condition.

The visitor would not find such prominent African amphibian groups as the caecilians, and the pipid and bufonid frogs. The advanced lizard families Agamidae and Lacertidae are missing and also the worm lizards belonging to the family Amphisbaenidae. The primitive, slender blind snakes (Leptotyphlopidae), the pythons (Pythoninae), the cobras and the coral snakes (Elapidae), and the vipers (Viperidae) are absent. An examination of the forest and savannah would fail to turn up characteristic African bird families such as the barbets (Capitonidae), honey-guides (Indicatoridae), trogons (Trogoniformes), hornbills (Bucerotidae), and secretary-birds (Sagatariidae). The primitive ratites, once a prominent part of the Madagascar bird fauna, are now extinct.

To the zoologist visitor, the most striking absence would be the great herds of large, herbivorous mammals and their predators; no antelopes, giraffes, zebras, rhinos, elephants, hippos, etc., and none of the big cats and wild dogs. In the trees there would be no monkeys and apes. Instead, he would find a small fauna in the same habitats; tiny shrews (Soricidae), tenrecs (Tenrecidae), and rodents (Cricetidae) being preyed upon by a variety of small, cat-like civits (Viverridae). The Madagascar trees would be occupied by four families of lemurs.

A look into the freshwater streams of Madagascar would reveal an almost completely different fish fauna. The ostariophysan fishes, comprising a multitude of families of catfishes, characins, and minnows — which dominate the freshwaters of Africa — are entirely absent. Absent also are many additional, old freshwater fish families that have probably been in Africa since the Mesozoic. In the Madagascar lakes and streams, their places are taken by euryhaline fishes, groups that can adjust their physiology to live in both high and low salinities. They can therefore migrate by entering the ocean. Examples are cichlids (Cichlidae), atherinids (Atherinidae), mullets (Mugilidae), gobies (Gobiidae), etc. Missing also in Madagascar waters are many groups of freshwater invertebrates.

The traveler would also find, in Madagascar, a depauperate fauna of terrestrial invertebrates. The termites, a relatively well known insect group, provide a good example. Africa has a large termite fauna and has probably served as the most important center of evolutionary radiation for that order (Isoptera). In Madagascar, the three largest families are well represented but the most primitive families, that have a predominantly southern hemisphere distribution, are missing. Also missing are an extensive group of genera that are considered to be the most advanced in the order. For some reason, Madagascar is host to the products of the middle stages of termite evolution.

If our hypothetical, newly arrived traveler from Africa was a botanist rather than a zoologist, he would gain an impression of a rich flora without many obvious gaps as far as major groups of plants are concerned. Probably the most noticeable differences on Madagascar would be in the degree of representation of various plant groups. He would notice the relatively high diversity among such groups as the orchids, composites, palms, and bamboos. Of course, the Didieraceae with strange cactus-like growth forms and several of the other endemic families would be conspicuous. At the generic and species levels, high percentages of endemism would be present.

Chapter 12

INDIA

We must recognize that it is abnormal conditions that account for much overseas dispersal. It is not the soft, gentle trade wind – it is the irresistible hurricane that is the key.

E.C. Zimmerman, *Pacific Basin Biogeography: A Summary Discussion,* 1963

When the breakup of Pangaea took place, India, of all the continents, had by far the longest journey, across the equator and well into the northern hemisphere. The pace of tectonic movement is deliberate so that it apparently took about 100 million years from the time that India – Madagascar separated from Africa until India made subaerial contact with Asia. Such an extended period of isolation should have produced a highly peculiar fauna and flora but Cretaceous – Paleocene fossils from India do not show this.

As as noted in the introduction to the biota of Africa, a worldwide, terrestrial vertebrate fauna existed during the Triassic period. However, in the case of India, Colbert (1973) noted that the large dinosaur faunas of the Jurassic and Cretaceous periods also indicated the presence of land connections to the rest of the world. Sullivan (1974) pointed out that three of India's late Cretaceous dinosaur species were strikingly like dinosaurs living then in South America. Most recently, Sahni (1984), who studied fossil material from the Cretaceous and Paleocene of peninsular India, found a general lack of endemism in the terrestrial faunas. Relationships at the generic and family levels were determined to lie with the Cretaceous vertebrates of Madagascar and Africa. In particular, the Upper Cretaceous dinosaurian fauna was similar including one common species.

Krommelbein (1979) has called attention to the significant geographical distribution of the Jurassic to Lower Cretaceous, marine ostracod genus *Majungaella*. Various species belonging to this genus have now been found in Tanzania, Madagascar, and northern India. This seems to be consistent with the observation of Krishnan (1974) who stated that the Jurassic system of India can be correlated almost zone by zone with the strata of Madagascar and East Africa. Krishnan also noted that the fauna of the Cretaceous system is better related to that of Madagascar, Natal, and West Australia than to the Tethyian region. The holostean fish genus *Lepidotes* has been taken from the Jurassic of India and Madagascar and also the Cretaceous of Africa and North and South America (Obruchev, 1964).

Some geological and geophysical data appear to indicate that India was originally attached to Antarctica and Australia and began to separate from them in the early Cretaceous (Johnson et al., 1976; Larson, 1977; Kennett, 1982). India, as it moved northward, was considered to have remained close to western Australia until about 105 Ma (mid-Cretaceous). Lillegraven et al. (1979) expressed the opinion that significant barriers to the exchange of land vertebrates from any continent with India existed from the early Cretaceous through the remainder of the Mesozoic. As India moved northward, the first collision was supposedly with an island arc lying seaward of a marginal basin below Asia during the Eocene and Oligocene (Curray et al., 1981). The main collision between India and the Asian continent has been stated to occur much later, during the Miocene (Gansser, 1964; Kennett, 1982). But a more recent analysis of magnetic anomalies in the Indian Ocean confirmed that the collision between India and Asia began about 50 Ma (Patriat and Achache, 1984). This early Eocene date is more consistent with the biological information.

The data on fossil plants is contradictory and, in large part, not consistent with that on the fossil vertebrates. Krassilov (1972) divided the Mesozoic (primarily Jurassic) world into three widespread floral provinces: a north temperate Phoenicopsis flora, a tropical Cycadeoidea assemblage, and a south temperate Pentaxyton group. Krassilov noted that the Mesozoic flora of India was of the Pentaxyton type and that it did not occur elsewhere in the northern hemisphere. He further observed that the fossil floras of peninsular India retained their southern affinity up to the Eocene and emphasized that the change in the floristic composition of the area coincided with its supposed collision with Asia.

On the other hand, Smiley (1979) maintained that the Jurassic and early Cretaceous floras of India are more closely related generically to the floras of Eurasia than to the floras of Gondwanaland. He said, for example, that 29 of 81 Jurassic genera (36%) occur on other Gondwana continents, in contrast to 48 genera (60%) that are present in the Jurassic floras of Eurasia; only 5 of the genera having affinities with Gondwanaland floras are restricted to Gondwana areas. Smiley also referred to a flora from Malaysia that was generically identical to one from northern India; the two floras were of the same age (near the Jurassic/Cretaceous boundary) and the species differences were slight.

Florin (1963) has shown that in the late Carboniferous to early Permian, the ancient conifer genera *Buriadia* and *Paramocladus* occurred in India and in South America. Another old conifer genus (*Walkomiella*) occurred in northern India and in southern Africa in the early Permian and in Australia in the late Permian. In the Podocarpaceae, the genus *Acmopyle* exists with two living species on new Caledonia and another on Fiji; fossil species have been found in the Eocene of South America and in the Oligocene of western Antarctica. The Jurassic genus *Retinosporites* of India is closely related to *Acmopyle*. Other podocarp genera, both living and fossil, occur in India but, since they also occur elsewhere in the East Asian area, their presence is not considered significant in terms of Mesozoic relationships. None of the large, widely distributed, northern genera of pines, firs, or cedars occurs on the Indian peninsula.

Toward the end of the Cretaceous, western and central India were covered by extensive sheets of lava called the Deccan Trap (Krishnan, 1974). In some places, the lava flows were as much as 1800 m in depth. This extensive volcanic activity, plus an infusion of African forms from the late Cretaceous connection, was undoubtedly responsible for the decimation of much of the earlier fauna and flora that India had transported from the south. Most of the remainder was probably eliminated as soon as India made effective terrestrial contact with Asia. One cannot imagine that the decimated remains of the original island biota would be able to compete with the diverse and relatively advanced biota of tropical Asia.

Despite a history of major volcanic disturbance and an early Tertiary invasion of Asian animals and plants, there remain in India some Tertiary fossil and recent indications of its past southern hemisphere relationships. In his discussion on the biogeography of the Indian peninsula, Mani (1974a) identified a number of phylogenetic relicts which he called "Gondwana faunal derivatives". These included the freshwater bivalve mollusc *Mulleria* which is also found in South America, two genera of land snails (Pulmonata), whipscorpions of the family Thelyphonidae, and five genera of gekkonid lizards (Gekkonidae). However, the latter three examples actually refer to groups that are primarily tropical, rather than south temperate, and some have virtually circum-tropical distributions.

The modern amphibian, reptile, and freshwater fish fauna of India has been reviewed by Jayaram (1974). Although the caecilians (Order Gymnophiona) are a very old amphibian group, they have essentially a circum-tropical distribution (Savage, 1973) so that presence in India does not necessarily demonstrate southern continent relationships. Fossils of a modern type of frog have been taken from the Eocene beds of India. But the relationships of the fossil (called *Indobatrachus*) are obscure. Savage suggested that it may be an early leptodactylid that had become isolated on India and had drifted northward with that island. The living families of Indian frogs (Bufonidae, Ranidae, Microhylidae, Pelobatidae, and Hylidae) probably represent post-Eocene invasions.

The Indian freshwater turtle fauna is essentially a modern Indo-Chinese assemblage. None of the old, southern hemisphere, pleurodiran forms are present (Jayaram, 1974). The lizard fauna consists of eight families with predominantly Indo-Chinese relationships. The one exception is the family Chamaeleontidae, a group that evidently evolved on Africa and Madagascar then spread north and eastward. It is considered to be an advanced family so that its spread to India probably took place in the late Tertiary. A few of the lizard genera of the mountainous regions show Palearctic relationships. The diverse snake fauna of India is almost entirely modern and Oriental in its relationships. No ancient relicts are to be found.

In regard to the bird fauna of India, ornithologists have recognized no important differences from the birds of the rest of the Oriental Region. As was mentioned in the discussion of African birds (p. 104), there is a close relationship between the two Regions with about 30% of the genera being shared (Welty, 1979) and most of the families. At the species level, however, the desert barrier formed by the Arabian Peninsula and the Sahara demonstrates its effectiveness since only 2% of the species are shared.

It is of considerable biogeographic interest to follow as closely as possible the sequence of changes that took place in India as that continent became fused with Asia in the early Tertiary. The contemporary mammalian fauna of India is well known and its relationships have been studied to the extent that it has been possible to reconstruct the major events that took place (Kurup, 1974). The history of the Tertiary fauna dispersal indicates that the invasions came through two great gateways, one at the Assam region to the east and one to the northwest. To the north, the rising Himalaya formed a great barrier so that, except for few montane species, entry was restricted to the two gateways.

The mammalian fauna of Assam (eastern India) is the richest and represents an almost purely Indo-Chinese association (Kurup, 1974). Of a total of 134 genera of land mammals in India, 85 (63%) are represented in Assam. A number of genera

(16) extend from Assam westward along a Himalayan strip that reaches Nepal but do not occur in peninsular India. Included are such forms as *Tupaia* (Tupaiidae), *Nyciticebus* (Lorisidae), *Hylobates* (Pongidae), and *Rhinoceros* (Rhinocerotidae). Many other mammalian genera demonstrate a discontinuous distribution, being present in Assam (and farther east) and the western Ghats toward the tip of the peninsula but entirely absent in the intervening area. In such cases, the species in the two areas are usually considered to be different. At the generic level, about 21% of the Assam fauna is considered to have entered from the northwestern gateway, either from the Ethiopian Region or from the Mediterranean portion of the Palearctic.

Whereas Assam received from the northwestern gateway about 21% of its mammalian genera, western India received about 36% of its genera from Indo-China (Kurup, 1974). This indicates that the flow of fauna from east to west was greater than in the opposite direction. Kurup concluded that the main eastern flow from the Indo-Chinese subregion entered India through Assam and bifurcated, one branch spreading to peninsular India and the other across the narrow, wooded sub-Himalayan belt to the northwestern parts and areas farther west. The other mainstream, formed of Ethiopian and Palearctic genera, entered India from the northwest and also bifurcated, one branch colonizing the peninsula and the other Assam.

Colbert (1973) noted that peninsular India is a blank as far as Paleocene to Pliocene land-living vertebrates are concerned. But, soon after, Sahni and Kumar (1974) reported a mammalian fauna of Middle Eocene age that exhibited strong Mongolian affinities. This discovery appears to be consistent with the mammalian invasion history worked out by Kurup (1974) on the basis of relationships found in the modern fauna.

Studies on the relationships of the freshwater invertebrate fauna of India are sparse and scattered. The freshwater snails of the family Potamiopsidae are thought by Davis (1979) to have originated in Gondwanaland and to have been transported to southeast Asia by the Indian plate. In this group, the more advanced genera are found in the Mekong River area while the three most primitive genera are found in South America, South Africa, and Australia. However, none of the older potamiopsid genera exists in India today and the Cretaceous fossils from there may not represent this group (the characters that identify the family are found in the snail rather than its shell). An aquatic oligochaete family, the Phreodrilidae, is confined to the temperate part of the southern hemisphere except for one northern locality and that is on the island of Sri-Lanka (Brinkhurst and Jamieson, 1971). However, this may not be particularly good evidence for continental drift because some of the species occur on the oceanic islands of the sub-Antarctic region – places that were probably completely glaciated then repopulated.

In general, the freshwater insects of India appear to comprise the more advanced members of their families and orders. There are no primitive stoneflies (Illies, 1968), mayflies (Edmunds, 1972), caddisflies (Ross, 1967) and other major groups. Singh (1974) discussed the local distribution of some of the aquatic Diptera. The mosquitos (Culicidae) have been studied because of their medical importance. There is a wet-season complex of mainly Indo-Chinese and Malayan elements that spawn during the retreating monsoon and a dry-season complex of mainly Ethiopian affinities with some Mediterranean influence.

As Lowe-McConnell (1975) has pointed out, the Indian freshwater fish fauna has many Malayan and south Chinese elements. Hora (1937) concluded that this fauna originated mainly in South China and spread westwards, aided by river captures and marshy conditions along the whole of northern India, some forms continuing westwards to Africa. There are some old fish fossils (the lungfish *Ceratodus* from the Upper Triassic and some ganoid fishes from the Upper Jurassic and Cretaceous) from India but they represent species that were not necessarily confined to freshwater. However, the Lower Eocene fossils of the families Osteoglossidae, Nandidae, Anabantidae, and Cyprinidae are significant. The latter three families have apparently always been confined to freshwater.

The family Nandidae (leaf fishes) is a small family with two genera in South America, one genus in Africa, and two in the Oriental Region; its distributional history is uncertain. However, the Anabantidae (climbing perches) and the Cyprinidae (minnows) are both best developed in southeast Asia and evidently originated there. They probably first entered Africa when a land connection across the Tethys Sea was established during the early Miocene (Briggs, 1979). Their presence in the Lower Eocene of peninsular India (Lowe-McConnell, 1975) undoubtedly indicates that India had, by that time, moved against the Eurasian continent. Since the Nandidae are in South America, they were also probably in Africa by the early Mesozoic and got across to South America, or vice versa, when the two continents were still in contact.

It is significant that all the freshwater fish genera common to India and Africa are also found in Asia east of India (Lowe-McConnell, 1975). At present, India has about 89 genera of primary freshwater fishes; of these, 66 also live in countries east of India, 19 both to the east and the west, and only one genus occurs only to the west. Also important is the fact that the 23 endemic Indian genera are all closely related to forms farther east (Menon, 1955). This information indicates that, beginning in the early Eocene, a major influx of freshwater fishes took place from the east and that many of the groups involved made their way across India and eventually to Africa by way of the Arabian Peninsula. Thus, Hora's (1937) original suggestion has been reinforced by more recent information.

The fish family Cichlidae is considered to be a secondary freshwater group since many of the species are quite euryhaline and can migrate across modest stretches of seawater. The family reaches its greatest development in Africa although it is also well represented in South and Central America. It also occurs on Madagascar and along the southern coast of the Indian Peninsula and on Sri-Lanka. Of the two Madagascan cichlid genera, one was apparently derived from African stock while the other is closely related to the Indian genus *Etroplus* (Lowe-McConnell, 1975). The latter genus probably represents an invasion from the west. Africa has evidently served as the major evolutionary center for the family and some cichlids extend northward to the eastern Mediterranean and the Jordan Valley.

Among the terrestrial invertebrates, the termites are probably the best known. According to Sen-Sarma (1974), the termites of India comprise predominately Oriental elements (65%). About 22% of the genera known so far are endemic. The Ethiopian elements come next in the order of importance and a few groups show Australian and Neotropical relationships. One living Indian species of the primitive family Termopsidae occurs in the mountains, mainly above an elevation of 1200 m. The harvester termites of the family Hodotermitidae are represented by the genus *Anacanthotermes*. This genus probably differentiated in Africa and then spread eastward. It ranges from North Africa and Eurasia south of the Caspian Sea to the western part of India.

The family Rhinotermitidae is well represented in India and some of the genera are broadly distributed within the tropics. The largest family of termites is the Termitidae. It contains nearly three fourths of the world's species. There are four sub-families recognized and all of them are found in India. Two of the genera, *Speculitermes* and *Synhamitermes* have peculiar disjunct distributions. The former is known from the Oriental Region and South America and the latter occurs only in India and South America. Sen-Sarma (1974) felt that these patterns must be explained on the basis of continental drift. However, the family itself is a large, modern group and one might question that the genera are really old enough to have been transported by continental movement. A better explanation may be that, considering Africa has probably been the major center of termite evolution (Emerson, 1955), these genera once occurred in Africa but have been competitively eliminated in that area leaving relict genera in the tropics of the New World and in southeast Asia.

The detailed biogeography of Indian butterflies (Lepidoptera) by Holloway (1974) is of considerable historic significance. He applied a cluster analysis method in order to identify the various faunal centers in the Old World tropics and in the Palearctic area to the north. By locating, in this manner, the older centers of generic development, he was able to trace the derivation of the modern Indian butterfly fauna. Holloway concluded that the colonization of southern India was primarily by species from the southeast Asian tropical centers, but with a few species derived from the African savannah regions via Arabia and the Middle East. The higher altitudes of the Himalaya, at or above the tree line, were colonized mainly by species from the Palearctic centers of the Pamirs and Turkestan.

The geography of the diplopod order Sphaerotheriida (giant pillmillipedes) has been discussed by Jeekel (1974). Within the family Sphaerotheriidae, the tribe Arthrosphaeriini occurs on the Indian Peninsula and Sri-Lanka while the Zoosphaeriini is confined to Madagascar. These two tribes are more closely related to each other than to any other sphaerotheriid group. Jeekel considered the order to be very old, possibly dating from the Paleozoic.

The general characteristics of the flora of India have been described by Mani (1974b). He noted that the richness of the flora is attributable to the immigration and colonization of plant species from widely different bordering territories. Though India is richer in plant species than perhaps any other equally large area in the world, there is a striking poverty of endemic genera. The Malayan floristic element is dominant, but there is also a large African element. There are Tibetan – Siberian elements in the alpine and higher zones of the Himalaya and Chinese – Japanese elements are evident in Burma and in the temperate zones of the southern parts of the eastern Himalaya.

INDIA SUMMARY

When India left the vicinity of its southern hemisphere neighbors and began its journey to the north, it was subjected to intense and widespread volcanic activity. This, plus an extended period of isolation, must have had a profound effect on its fauna and flora. Thus, when contact was made with northeastern Africa in the late Cretaceous and then with Asia in the lower Eocene, an impoverished and weakened biota was probably quickly overwhelmed by continental invaders. Consequently, there are almost no living relics that one can confidently trace back to the Mesozoic phase of Indian history.

In looking for evidence of relationship among the early fossils reported from India, one can find several helpful indications. In the late Carboniferous to early Permian, two conifer genera have been found in India and in South America; another conifer genus occurred in the early Permian of India and southern Africa; and an Indian Jurassic conifer genus is closely related to one that was widespread in the southern hemisphere (Florin, 1963). Krassilov (1972) noted that a distinct, south temperate, pentaxyton flora occurred in India in the Mesozoic. But Smiley (1979) maintained that the Jurassic and early Cretaceous floras of India were most closely related to those of Eurasia.

The observation of Krommelbein (1979) that a marine ostracod fauna of the Indian Jurassic/Lower Cretaceous was very similar to one found in Madagascar and Tanzania seems to correlate well with the statement by Krishnan (1974) that the Jurassic system of India can be matched almost zone by zone with the strata of Madagascar and East Africa. On one hand, some geological and geophysical data appear to indicate that India began to separate from its Gondwana connections in the early Cretaceous (Johnson et al., 1976; Larson, 1977; Kennett, 1982) yet, on the other hand, the late Cretaceous dinosaurs of India appear to closely resemble those from other parts of the southern hemisphere (Colbert, 1973; Sullivan, 1974) and the Cretaceous/Paleocene fauna shows a general lack of endemism (Sahni, 1984).

When did India make contact to form a continuous land mass with Asia? Tertiary fossils from India are scarce but the remains of two families of freshwater fishes discovered in Lower Eocene deposits (Lowe-McConnell, 1975) are particularly important. The Anabantidae (climbing perches) and the Cyprinidae (minnows) apparently evolved in southeast Asia and, although they are in Africa today, they were evidently unable to enter that continent until the early Miocene. Therefore, their presence in India must be attributed to an invasion via freshwater stream capture after India became a part of Asia.

An analysis of the Indian freshwater fish fauna as a whole (Lowe-McConnell, 1975) has shown it to be a very rich assemblage with strong relationships to the eastern part of the Oriental Region. The great majority of the Indian genera are closely related to forms farther east. These relationships provide a scenario of events that probably took place rapidly following the early Eocene docking of the Indian island. A major invasion of a modern, diverse freshwater fish fauna took place from the east via the Assam gateway. There may have also been an invasion of Ethiopian stocks from the northwest but this was probably limited to the Cichlidae and the Cyprinodontidae, saltwater-tolerant families that could have migrated from Africa along the coastline.

The local distributions and relationships of two groups of Indian terrestrial animals have been investigated closely enough to provide an account of the invasion events that took place in the terrestrial environment. Among the mammals, the Indo-Chinese influence is predominant although, in western India, there is a strong African influence (Kurup, 1974). The flow from the east entered through Assam and bifurcated, one branch spreading to peninsular India and the other across the sub-Himalayan belt. Another stream, comprised of both Ethiopian and Palearctic genera, also bifurcated, one branch colonizing the peninsula and the other extending directly eastward. Fossil mammals from the Middle Eocene were shown to have Asiatic relationships.

The biogeography of the Indian butterflies by Holloway (1974) gave results similar to those reported for the mammals. The colonization of the peninsula was primarily by species from the southeast Asian tropical centers but with a few species derived from the African savannah regions. The higher altitudes of the Himalaya were colonized mainly by species from Palearctic centers. The general relationship of the flora, as described by Mani (1974b), corresponded remarkably well with the general scheme provided by the mammals and butterflies.

Chapter 13

SOUTHERN CONTINENTS SUMMARY

> As plates and plate boundaries multiply, small platelets appear and disappear for little reason other than that without them a postulated set of plate movements is not feasible. Reconstructions of plate history have become much more uncertain, diverse, and above all, idiosyncratic.

> Tjeerd H. van Andel, *New Views on an Old Planet*, 1985

Unlike those of the north, the faunas and floras of the southern continents have, for the most part, been well separated since the early Cretaceous. These separations have allowed natural selection to produce extensive evolutionary changes that are intrinsically interesting to the biologist. More important to the biogeographer, is the fact that the evolutionary changes reflect the history of the geographical relationships. In the southern hemisphere especially, the historical geology is not well known. As our knowledge of present and past biological relationships increases so will our understanding of the geological events. It will not be possible to construct a thorough account of the history of the southern hemisphere without correlating the evidence from both the biological and the earth sciences.

NEW ZEALAND

It seems apparent that the rich continental biota that existed during the early Mesozoic did not extend to New Zealand. Particularly noticeable by its absence is the diverse Triassic and Jurassic vertebrate fauna including amphibians, reptiles, early mammals, and freshwater fishes. The fact that *Sphenodon* and *Leiopelma* are present does not require a land connection. *Sphenodon* is a lizard-like reptile and we know that some lizard families (Gekkonidae and Scincidae for example) are remarkably good at overseas dispersal. The case of *Leiopelma* is not unique for other frog genera (*Rana* for example) have managed to cross significant saltwater barriers. The fossil remains of a late Cretaceous dinosaur are a puzzle. It might have been an aquatic creature that had crossed a considerable sea barrier.

On the other hand, the biological data to indicate that New Zealand must have undergone considerable movement in relation to the southern continents. The old ties to southern South America are particularly noticeable and indicate that New Zealand was in a geographic position to receive considerable input from South America either directly or via Antarctica. Furthermore, it appears that most of this relationship was not mediated via Australia. There is very little evidence of any Mesozoic relationship to Africa or India other than fossils of a few groups that were widespread such as the conifer genera *Podocarpus* and *Araucaria*. The relationship to Australia, although continuous to some extent, appears to have developed more strongly beginning in the mid-Tertiary as if the two areas had moved closer together at that time. There is a marked and special relationship to New Caledonia but not to the extent that New Zealand and New Caledonia could have been a single land mass.

The biological data suggest that New Zealand may have occupied a position considerably south and west of its present location. If it were as much as 20° south and 10° west, it could have picked up its old biota from an unglaciated Antarctica (or a separated western Antarctica) that, in turn, was joined by a peninsula or an archipelago (the Palmer Peninsula plus the islands of the Scotia Arc) to the tip of South America. Thus, a rich South American biota would have been reduced by a filter bridge to Antarctica and then further selected by a significant water gap before reaching New Zealand.

AUSTRALIA

On the whole, the biological evidence does not appear to support the concept of a post-Triassic land connection between Australia and Antarctica. In all probability, Australia received most of its ancient fauna via waif dispersal from Antarctica over some oceanic expanse. At the same time, it was not so isolated from the East Indies that it was prevented from picking up much of its ancient flora from southeast Asia via islands or archipelagos. It appears that a rift zone began to open between Australia and Antarctica in the late Jurassic and that deep-sea conditions extended along most of the rift by 80 Ma. Although most geophysical reconstructions do not show a complete separation until the Paleocene or Eocene, the biological data do not require a connection since the early Mesozoic. For terrestrial animals and plants, an epicontinental sea provides a barrier just as effective as the deep sea.

The southern land mass relationships of Australia are, for the most part, quite separate from those of New Zealand, especially where the ancient fauna is concerned. Australian groups such as the frogs, turtles, land reptiles, birds, and mammals that were evidently received from South America via Antarctica are not related to the vertebrates that were received from the same source by New Zealand. The New Zealand tuatara and leiopelmatid frogs are older vertebrate types and not akin to anything in Australia. The New Zealand ratite birds belong to a different order and New Zealand received no freshwater turtles or mammals. New Zealand possesses two lizard families (Gekkonidae and Scincidae) but they probably reached there in the mid-Tertiary or later. These biological relationships appear to reflect the fact that Australia and New Zealand were close to Antarctica at different times. As New Zealand and then Australia moved away from Antarctica, they probably approached each other, permitting the late Tertiary and Quaternary migrations that have caused the modern New Zealand biota to resemble that of Australia.

It is important to note that the Mesozoic movement of living organisms between Australia and South America via Antarctica was not entirely unidirectional. It is likely that southern South America received much of its primitive flora such as liverworts, *Podocarpus,* and *Nothofagus* from Australia. Some of the faunal elements, well established in South America today, probably originated in Australia. Examples are parrots, pigeons, and the parastacid crayfishes. Many other groups of the old fauna indicate Australian – South American relationships but the direction of migration is not easily determined.

ANTARCTICA

The early Mesozoic fossils found in Antarctica indicate that it provided a suitable habitat for many elements of the worldwide, interrelated vertebrate fauna that existed in the Triassic. At the same time, there existed a diverse, primitive flora that was gradually replaced toward the end of the Mesozoic by flowering plants and conifers.

The discovery of late Eocene marsupial remains served to emphasize the role of the Antarctic as a connecting link between South America and Australia. Since the Antarctic specimens closely resemble taxa that lived in South America about 50 Ma, it was suggested that the migration took place at about that time. This estimate is consistent with the probable age of the Australian marsupial fauna.

When the fossil evidence is added to the reported patterns of contemporaneous relationships, the central role of the Antarctic in southern hemisphere biogeography becomes clear. These relationships became established over a very long period of time (throughout the Mesozoic and early Tertiary) and involved a broad spectrum of the flora and fauna. In an evolutionary sense, the continent may have served as center of origin for some of the southern hemisphere biota that are now found in peripheral locations.

SOUTH AMERICA

There is no doubt that, in the Triassic, South America was closely joined to the other continents. In fact, at this time, its vertebrate fauna was closer to that of Africa and Europe rather than North America. However, by the end of the Jurassic, with the development of the South Atlantic Ocean, this relationship came to an end. Since there was probably no direct mid-Mesozoic connection to North America, the separation from Africa resulted in an isolation that was to continue for the remainder of the Mesozoic and for the entire Tertiary. Consequently, South America developed a unique and highly peculiar biota. This is particularly evident in groups such as the birds, mammals, and angiosperm plants which have undergone most of their evolution in post-Jurassic times.

The most convincing evidence of a mid-Mesozoic relationship to Africa lies in the distribution of the freshwater fauna. In Upper Jurassic times, as indicated by the freshwater ostracods and ostariophysan fishes (and their nematode parasites), there was probably a peninsular connection between the northeast corner of Brazil and the African coast in the Gulf of Guinea. This connection may have been low and swampy. It permitted some of the African ostariophysans to gain access to South America and a few other freshwater fishes (Osteoglossidae, Nandidae) may have also entered. The traffic in the opposite direction probably included various groups of amphibians (caecilians, frogs) and some freshwater turtles and insects. Other old groups such as the crayfishes, freshwater crabs, and water mites may have gotten across in one or both directions. In contrast to the good evidence for the mid-Mesozoic exchange of freshwater faunas, the old terrestrial groups such as the land snails and some of the insects do not seem to be as closely related to the African forms. In fact, most of the old aquatic and terrestrial invertebrates and the most primitive plants show a stronger affinity to Australia than to Africa. These differential patterns, along with our improved knowledge of the shallow-water marine faunas of the South Atlantic, allow us to provide useful biological data for the drift that took place between South America and Africa.

In regard to the possible connection of South America to Antarctica, none of the exchanges with Australia, with the possible exception of the Triassic amphibians and reptiles, would have absolutely required such a bridge. As Simpson (1980) has stated, "South America may then also have been closer to Antarctica than at present, but that is not made necessary or even clearly probable by any known fact." A larger Antarctic (without the burden of an ice pack) covered with vegetation could, in its present position, have acted as a potent agent for the distribution of the southern hemisphere biota.

Although South America has received many groups of its biota from external sources such as Australia, Africa, and North America, it has also served as a center of evolutionary origin for a number of groups of animals and plants that subsequently invaded other continents. In the herpetofauna, two frog families, two freshwater turtle families, a lizard family, and a crocodile were probably exported to Australia. Herpetofauna donations to Africa probably include the caecilians, two frog families, and a lizard family. Two South American bird families plus a variety of amphibians and reptiles have invaded North America, marsupial mammals reached Australia, notoungulate and edentate placental mammals and marsupials reached North America and Eurasia, water striders and bees have become very widespread, and some of the higher plant families have reached other southern continents as well as North America.

AFRICA

The general relationships between Africa and South America have largely been summarized in the foregoing account. However, when one examines this relationship in detail, it may be seen that Africa has given more than it has received. Among the major groups that have reached South America from Africa are probably three frog families, a lizard family, a primate stem group (Omomyidae via North America) from which the New World monkeys may have been derived, ostariophysan fishes (characins and catfishes) and their nematode parasites, termites, and ants.

The biological ties between Africa and Eurasia are much stronger than those between Africa and South America. At various times during the Tertiary, there was evidently intermittent contact across the Tethys Sea. This relationship is well illustrated by the mammals because the fossil evidence is better. In the Paleocene and Eocene, it is likely that at least three primitive mammalian groups reached Africa from the north. The most dramatic faunal change occurred in the late Oligocene/early Miocene when 29 new families and 79 new genera made their appearance. Archaic suids, canids, viverrids, and felids arrived from Eurasia. By the late Miocene, another 18 families appeared. The Pliocene and Pleistocene saw more new groups arrive in Africa but these times were also important for the export of African mammals to Eurasia.

Compared to South America, the southern continent relationships of Africa are weak. While it is true that there are some examples of Gondwana affinities such as certain groups of aquatic insects and molluscs, a family of frogs, and a termite genus, they are few. The flora also does not indicate much in the way of southern ties. Relict conifer and taxad genera that are numerous in Australasia and South America are notably absent in Africa. This situation plus the evident close proximity of Africa to Europe and Asia that existed throughout the Tertiary, makes one suspect that there never was a direct Mesozoic connection between Africa and Antarctica.

In the Introduction (p. 59) it was noted that the Lower Triassic Middle and Upper Beaufort beds in Africa had yielded fossils that are named for their characteristic reptiles, the *Lystrosaurus* zone and the *Cynognathus* zone. When the same faunas were discovered in the Fremount Formation of Antarctica, it was assumed that here was proof that Africa and Antarctica had been joined. However, it has since become apparent that these faunas were very widespread occurring in China, European Russia, and South America (at least the *Cynognathus* zone). Taking into consideration the generally closer Mesozoic relationship between South America and Australia – New Zealand, it seems reasonable to propose that the Antarctic fossils came from South America rather than Africa. This would allow one Triassic pathway or filter bridge to the Antarctic and permit both the Antarctic and Africa to remain closer to their present latitudinal positions.

MADAGASCAR

To the biogeographer, the most striking thing about Madagascar is its very un-balanced or disharmonic fauna, reflecting an extended isolation, and the fact that those forms that are present are not extremely ancient. Although there is nothing in the present biota that could not have reached the island over a seawater gap, the fossil evidence indicates a much closer past relationship to Africa. The earliest fossil reptiles recovered from deposits on Madagascar came from the Upper Permian. Three widespread families are represented and it is interesting to see that all three are adapted for an aquatic existence. Later, from the Middle Jurassic to the Upper Cretaceous, Madagascar was occupied by dinosaurs without obvious aquatic adaptations.

It has been noted that the earliest ostariophysan fishes would have needed to enter Africa by the Upper Jurassic in order to take advantage of the remaining connection to South America. These primary freshwater fishes have been able to dominate all of the aquatic environments to which they have been able to gain access. It seems reasonable to assume that if a substantial connection to Madagascar still existed in the Upper Jurassic, the ostariophysan fishes would have taken advantage of it and would be in Madagascar today.

It is known that the terrestrial tetrapods of the Triassic period had an essentially worldwide distribution including Madagascar. The Upper Jurassic dinosaur faunas of the northern and southern hemispheres remained similar but most of the new dinosaur families that evolved in the Cretaceous of the north failed to reach the southern hemisphere. It should also be noted that Triassic fossils of an ancient conifer (*Voltziopsis*) have been reported from both Madagascar and Africa. This information appears to place some time constraints on the relationship between Madagascar and Africa. It suggests a scenario that would have Madagascar approach close to Africa in the Upper Permian, effect an actual connection in the Triassic, and drift away again by the Upper Jurassic.

The final important fact about the biological relationships of Madagascar is that there are very few phylogenetic relicts that point to ancient Gondwanaland or Antarctic ties. It has been noted that the theory of a Gondwana origin for the ratite birds, including the elephant birds of Madagascar, is questionable. Some of the freshwater insect fauna, such as a few genera of Diptera, Trichoptera, Ephemeroptera, and Coleoptera, and a freshwater snail genus, appear to have southern continent relationships. One can also find a few examples among the terrestrial insects. In the flora, there is one species of a very primitive angiosperm and one genus (with six species) of an old gymnosperm, both of which may have been derived from the south. However, in general, such Gondwana relationships are weak.

INDIA

The details of the physical and biological relationships of India to Madagascar and Africa constitute the greatest puzzle of the southern hemisphere. Supposedly, Madagascar – India became separated from east Africa in about mid-Jurassic times, some 150 Ma (Rabinowitz et al., 1983). This was determined on the basis of geophysical evidence but it is consistent with the geological and biological data that are currently available. It has been suggested (p. 160) that the break between Madagascar and India occurred considerably later (early Cretaceous). Since India did not establish a subaerial connection to Asia until the early Eocene, about 50 Ma, this means that India before its docking should have demonstrated in its biota the effects of about 100 Ma of isolation from Africa.

However, the Cretaceous/Paleocene fauna of India, which is only now becoming better known, does not reflect an extended isolation at all (Sahni, 1984). The vertebrate fossils demonstrate a close relationship at the generic and family level to the faunas of Madagascar and Africa. One Upper Cretaceous dinosaur species was found to be common to all three areas. Considering such evidence, Sahni predicted the presence of a filter corridor during the Upper Cretaceous, about 80 Ma. What form could such a corridor have taken? Plate tectonic maps for that time (Barron et al., 1981; Kennett, 1982) show India sitting by itself far out in the Indian Ocean.

India apparently underwent most of its northward movement from about 90 to 50 Ma. The Ninetyeast Ridge and the Chagos-Laccadive Ridge have been considered as resembling train tracks upon which India traveled on its journey toward Asia (Kennett, 1982). This is not a good analogy. While these north-south ridges may mark to some degree the passage of India, there is no indication that they stretched ahead of that continent as it moved. A more westerly feature of the Indian Ocean sea-floor topography is the Owen Fracture Zone, a north-south ridge that runs close to the tip of the Somali Peninsula.

If, in its northward journey, India stayed closer to the East African coast than is depicted in the geophysical reconstructions, and if the Owen Fracture Zone marked the passage of its western margin, then a connection to Africa in the vicinity of the Somali Peninsula appears to be plausible (Map 5). This hypothesis may, at the same time, explain the position of the Seychelles platform which could have broken off from India as that continent moved northward. As was noted earlier (p. 123), the fossil and stratigraphic evidence indicates that in the Triassic India was attached to Madagascar, and through it to east Africa, rather than to Antarctica and Australia.

It seems apparent that the late Cretaceous relationship of India to Africa, which the fossil materials indicate, was attributable to a direct connection rather than via Madagascar. Sahni (1984) referred to the presence in India of Cretaceous/Paleocene pelobatid frogs and anguid lizards. These are holarctic forms unknown in Madagascar but they could have been present in northeast Africa as the result of intermittent contacts between that continent and Europe. Also, considering that India probably began its northward movement about 90 Ma and ceased it about 50 Ma, it could have been opposite and close to northeast Africa some 70 – 80 Ma.

While the recognition of a new land bridge should not be undertaken lightly, in this case, there seems to be no reasonable alternative. There is no way in which India could exhibit such a close biological relationship to Africa and, at the same time, have been isolated from it for tens of millions of years. Sahni (1984) suggested a possible filter corridor involving the Mascarene Plateau and the Chagos-Laccadive Ridge but these features are too far east. The hypothetical position of India during the late Cretaceous/Paleocene and the presence of certain northern genera seem to favor a direct connection to northeastern Africa.

Among the few living relics that might extend back to the pre-drift period of Indian history are the bivalve mollusc *Mullaria,* two genera of land snails, an earthworm species of the family Pheodrilidae, and some millipeds of the family Sphaerotheriidae. In general, it may be observed that the historical biology of India indicates an affinity with Madagascar and East Africa as opposed to Antarctica and Australia. At the same time, it should be noted that evidence of a general temperate, southern hemisphere relationship is weak. The Mesozoic fossil record is reasonably good but many typical Gondwana elements are missing. This is probably a reflection of the fact that Africa itself does not have strong Gondwana connections.

Part 3

THE OCEANS

Chapter 14

THE OCEANIC PLATES

> No matter how strange or rigorous the inanimate or noncompetitive environment may be, the evolutionary products of such environments are not as thoroughly refined by competition, and thus are not as well prepared for widespread success, as the products of highly competitive associations.
>
> Hobart M. Smith, *Evolution of Chordate Structure,* 1960

In any evaluation of biogeography that is conducted on a global scale, one must bear in mind that the oceans cover 71% of the earth's surface. Compared to the atmosphere, the fluid of the hydrosphere is viscous enough to provide physical support and contains sufficient nutrients so that almost all its volume is continuously occupied by a vast array of living things. Here, we find that life exists on a truly three-dimensional scale with its vertical component extending from the surface to the greatest depths – almost 11,000 m.

Although the shapes of the major terrestrial biogeographic regions were determined by Sclater (1858) and Wallace (1876) over a hundred years ago, it has only been recently that a general agreement has been reached concerning the patterns of life in the surface waters of the oceans (Briggs, 1974). The hydrosphere may be apportioned into four temperature zones and the biogeographic regions occupy parts, or sometimes all, of these zones. Since the major oceanic gyres have an effect on the distribution of surface temperatures, the shapes of the zones are influenced accordingly (Fig. 14).

In comparison to the shallow waters of the continental shelves and of the epipelagic zone of the open ocean, the horizontal distribution of life at greater depths is very poorly known. However, sufficient sampling has been done at various depths so we have dependable information about the vertical distribution of marine life. For example, distinct biotic communities have been described from the mesopelagic, bathypelagic, and hadopelagic zones of the open ocean. Similar, depth-related changes are known to occur in the benthic realm. These vertical divisions are usually found to take place at similar depths throughout the world ocean (Fig. 15).

As on land, plate tectonic activity in the oceans has had profound biological consequences. The continental dispersal that began in the late Triassic took place in conjunction with, and may have been the result of, renewed activity among the lithospheric plates forming the earth's crust. It has been observed that such plates interact in three ways: they may diverge, converge, or slide past one another (Kennett, 1982).

When plates diverge from one another they allow hot mantle material to upwell to the surface and to form new sea floor. As the divergence takes place, the upwelled material forms prominent ridges that often extend thousands of meters above the older sea floor. The ridges are generally formed at about 90° to the two diverging plates. The elevation of the ridges is due to the fact that they consist of rock that is hotter and less dense than the surrounding sea floor.

It has been suggested that the faster the divergence, the greater the width of the ridges (Fischer, 1984). Thus, in times of high plate activity many broad ridges will be produced. These will, in turn, restrict the volume of the ocean basins and cause the formation of epicontinental seas. As the hot ridge material gradually moves in a lateral direction, it begins to cool and contract. As contraction takes place, the water depth increases. For example, the Mid-Atlantic Ridge lies about 2500 m below sea level. By the time the new crust produced at the ridge is about 2 million years old, the depth will have increased to about 3000 m. When the crust ages to about 20 million years, the depth will be about 4000 m and at 50 million years it will be at about 5000 m (Sclater and Tapscott, 1979). In this manner the ridges form prominent, elongate elevations that tend to separate the various oceanic basins.

In addition to separating the oceanic basins, the ridges have another significant function in a biogeographic sense. Seawater that has seeped into the fractured rock of the ridges is often expelled in hydrothermic vents. Reduced sulphur compounds are emitted from the vents and are utilized by chemosynthetic bacteria. The bacterial production provides a food base for a complex community. Many of the species of such communities were unknown before the vents were discovered and may be confined to the vent habitats (Rona et al., 1983).

When two plates converge, one is usually gradually destroyed by being subducted beneath the other. The downward movement of the lithosphere as it is being sub-

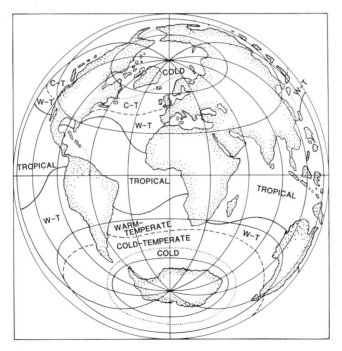

Fig. 14. The four major temperature zones of the world ocean. The various marine biogeographic regions are located within the zones. The lines separating the zones depict areas of sharp latitudinal change in the biota. After Briggs (1974).

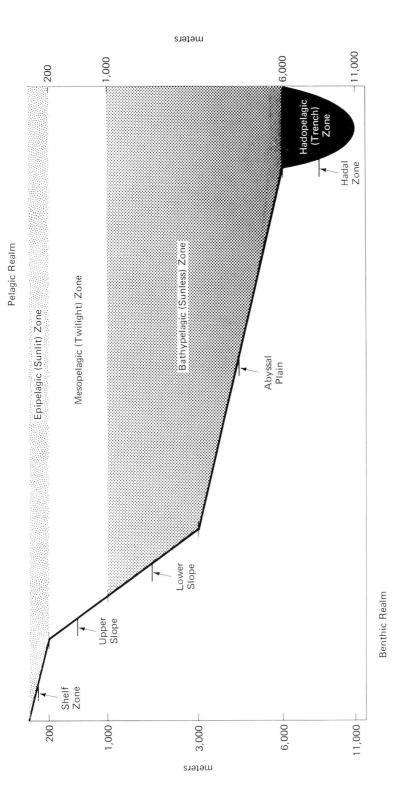

Fig. 15. Diagrammatic representation of the vertical divisions of the world ocean based on the distribution of the biota. After Briggs (1974).

ducted is responsible for the formation of the oceanic trenches. Trenches may occur as the result of the convergence of two oceanic plates as in the formation of the Kermadec-Tonga Trench or the convergence of an oceanic and a continental plate as in the case of the Peru-Chile Trench. Trench formation is usually accompanied by volcanic activity which, in the oceanic environment, may result in the formation of an island arc. Depending on the location, an ocean basin may be isolated behind the island arc. The trenches may extend as much as 5000 m below the abyssal plain.

When oceanic plates slide past one another, they form transverse fractures or transform faults. Since these structures are slip lines and generally do not involve the formation or destruction of crustal material, they are considered to be conservative plate margins (Kennett, 1982). Also, the mid-ocean ridges often become offset by transform faults which serve to accommodate the differing lateral motions of the diverging plates. The Gulf of California is part of a system of transform faults. Such structures may have a notable effect on the distribution of the marine biota so can be important in the biogeography of certain areas.

THE OCEAN BASINS

As Murray (1895) suggested long ago, it is possible that the deep seas were anaerobic during most or all of the Mesozoic. Vertical circulation may have been prevented by the presence of relatively warm surface waters everywhere so that multicellular abyssal life could not exist until the poles became cold and the dense, oxygenated water could sink into the depths. Regardless of whether a Mesozoic deep-sea fauna was decimated by falling temperature at the end of that era or whether such a fauna existed at all, the present fauna of the great depths seems to be a relatively new one. The present deep, cold-temperature regime $(1 - 2.5°C)$ may not have become established until the late Pliocene so it is possible that most of the invasions into the abyssal and trench habitats have taken place since that time (Briggs, 1974).

Considering the uniformity of the physical surroundings in the deep basins, the presence of a soft mud substrate almost everywhere, and the predominance of a single feeding type (deposit feeders) among the animals, one might conclude that species – once having become adapted to such an environment – would become exceedingly widespread. The early naturalists did expect to find a monotonous, wide-ranging abyssal fauna and, as recently as 1957, Bruun predicted that the future would show that in general the true abyssal species are cosmopolitan because the abyssal zone formed an ecological unit with no barriers except for isolated areas like the Sulu Sea.

As our knowledge of sea-floor topography and the ocean-basin fauna increased, it became obvious that there were substantial geographic differences. Clarke (1962), in his review of the abyssal molluscs of the world, indicated that a high degree of endemism exists with respect to the individual basins. He listed a total of 1087 species (not including cephalopods) that are found below 1800 m and noted that, although there are 47 recognized ocean basins, the mean distributional spread was only about 2.0 basins per species. In nearly 90% of the cases where more than one basin was occupied by a species, the basins proved to be contiguous. Menzies (1965) compared the abyssal isopods from the adjacent Argentine and Cape Basins in the South Atlantic: of a total of 58 species, only 14% occurred in both areas and none of the species were known from the other oceans.

Vinogradova (1979) examined the geographical distribution of the abyssal and hadal fauna from a worldwide perspective. She called attention to the fact that the degree of taxonomic isolation (endemism) of the deep-sea fauna increases with depth. For the world ocean, about 85% of the deep-sea fauna was found to occur in one ocean only and, only 4% turned out to be common to the Pacific, Indian, and Atlantic Oceans. It has been noted that the fishes, most molluscs, isopods, amphipods, ascidians, and sponges appear to be highly restricted in their horizontal distribution, while the echinoderms, hydroids, and bivalve molluscs have more broadly distributed species (Briggs, 1974). Perhaps, after the abyss achieved its present cold temperature, those groups with species that are now most restricted moved in first while those with wide-ranging species may be comparatively recent arrivals.

In regard to the deep-sea trenches which comprise the Hadal Zone, the animals inhabiting such depths have become highly distinct. The most comprehensive analysis of this fauna was done by Belyaev in 1966 (with an English translation being published in 1972). He found that, in general, some 64% of the species and about 10% of the genera taken below 6000 m were endemic to such depths. Even more important, each deep-sea trench was found to contain a high proportion of endemic species, an average of 58%. An analysis of the bivalve mollusc fauna by Knudsen (1970) showed that all Hadal Zone specimens were apparently endemics.

When we consider that the abyssal basins tend to be separated from one another by oceanic ridges and rises that have resulted from plate divergence and that the trenches and island arc basins are the result of plate subduction, the profound effects of plate tectonics on the fauna of the deep sea becomes apparent. Certainly, the greater part of the species diversity that is found in the modern deep sea is attributable to the barriers erected and the dispersal opportunities afforded by the gradual movement of the plates that comprise the crust of the sea floor.

THE SHALLOW OCEAN

In studying the long-term effects of plate movement on the biota of the continental shelves, we tend to focus on the rather obvious effects of continental separations and joinings. In Chapter 1 attention was called to Fallow's (1979) work on the marine fauna of the continental shelf on each side of the North Atlantic. He found a strong, positive correlation between the width of the ocean basin and the degree of similarity of the invertebrate animals that inhabited the shelf on each side. As the North Atlantic became gradually wider, gene flow between east and west became reduced and the resulting evolutionary changes became greater. Similar effects have been noted in regard to the separation of Africa from Caribbean – South America and the departure of Australia from the vicinity of Antarctica.

In cases where continents have closed with one another, creating terrestrial connections, the effects on the marine biota have been rapid and dramatic. Examples are the disruption of the Tethys Sea in the Miocene by the connection between Africa and Asia and the separation of the Bering Sea – Arctic Ocean biota by the Bering Land Bridge. The severing of the New World tropics by the Isthmus of Panama was due to volcanic activity stimulated by subduction along the Mexico Trench. In each of these examples, the separation put an immediate stop to gene flow and the divided populations became almost completely distinct at the species level within a few million years. Aside from the separation and joining of the continents, plate tectonics has affected the distribution of shallow-water marine organisms in one other important way. This is in the creation and subsidence of oceanic islands.

Almost all oceanic islands, that are separated from the nearest mainland by extensive stretches of deep water, are volcanic in origin. Furthermore, as Wilson (1963) first showed, there is an interesting age relationship between many islands and the plate margins as identified by the mid-ocean ridges. In the Atlantic Ocean for example, Ascension Island lies very close to the ridge and is only about 1 million years old. The Azores, St. Helena, and Gough Islands are somewhat removed from the ridge and are about 20 million years old. Finally, such islands as Bermuda, Canaries, and Cape Verde are still farther away and are much older. This distance-age pattern caused Wilson to suggest that such islands originate along the ridges where volcanic activity is the greatest. Once such an island was formed, it would be carried laterally as new sea floor was created at the ridge. However, as the sea floor moved away from the ridge, it became cooler and sunk to greater depths. This often caused the islands to sink beneath the surface to form sea mounts or guyots. The larger islands would tend to persist longer but eventually all would be consumed in the trenches.

Wilson's (1963) observations were compatible with and helped to support the (then) new theory of sea-floor spreading. Wilson (1973) also proposed a mechanism for the formation of certain island chains located far from plate margins. These appear to have arisen as the result of static "hot spot" activity. As a plate gradually moves over such a hot spot or magma source, a string of volcanic islands may be produced. The chain of islands comprising the Hawaiian archipelago is a good example. The hot spot currently lies under the big island of Hawaii and the islands to the northwest become progressively older with distance from Hawaii. Other examples of island chains formed as the result of hot spot activity are the Society, Marquesas, and Galapagos in the Pacific, Tristan da Cunha and Bouvet in the Atlantic, and the Prince Edward – Marion and the St. Paul's – Amsterdam groups in the Indian Ocean (Kennett, 1982).

Island chains, once formed, may be transported from their original positions along with the movement of the plate itself. In the Pliocene, there occurred a rather sudden invasion of Indo-West Pacific hermatypic corals into the eastern Pacific (Dana, 1975). Dana suggested that this invasion was the result of the Pacific plate having moved the Line Islands into the path of the Equatorial Countercurrent, that current then being able to transport the coral larvae eastward across East Pacific deep-water barrier.

When plate subduction occurs in the oceanic environment, it often results in the formation of three related structures: island arcs, trenches, and back-arc basins. The latter two have been discussed but island arcs form very important structures, particularly in the Pacific Ocean. Some of the island arcs occur in isolated locations far from continents or large islands. Others are continuous with or close to terrestrial areas or their continental shelves. Examples of isolated arcs are Kermadec-Tonga, New Hebrides, and the Marianas. Arcs closer to the mainland are the Kuriles, the Aleutians, and the Lesser Antilles. The relative geographic isolation of such arcs have an important bearing on their biogeographic significance.

Another, rare, category of oceanic islands are those that are apparently formed from remnants of continental crust. The Seychelles comprise the best known example. The platform from which they project was probably left behind as India separated from Madagascar and began its northward journey (Braithwaite, 1984). In the terrestrial flora there is one monotypic family, the Medusagynaceae, and nine endemic genera (Procter, 1984). Among the terrestrial animals, the amphibia are the most distinctive with a monotypic family of frogs, the Sooglossidae, and three endemic genera of caecilians (Nussbaum, 1984). Another possible example is New Caledonia with its strange endemic families and genera of plants (Holloway, 1979).

Oceanic islands and island chains that are well isolated (300 miles or more from the nearest mainland or island group) and are of sufficient age (Pliocene or older) tend to demonstrate in their terrestrial and shallow-water marine biota remarkable evolutionary changes. Once an isolated island or island group is established and, by means of fortuitous accumulation picks up its founder species, the various populations can be expected to embark immediately on their own lines of evolutionary change. Since relatively small populations are usually involved and because many aspects of the ecology are apt to be different, it may expected that such change would occur rapidly in comparison to a mainland situation. As a result, oceanic islands that are relatively old possess faunas and floras that show a high degree of evolutionary divergence. The best indication of such divergence is the rate of endemism (Briggs, 1966).

Isolated oceanic islands with known ages are of great interest to the evolutionary biologist for they represent natural laboratories where the results of experiments in natural selection, competition, behavior, and ecology can be examined. Hubbell (1968) pointed out that isolated archipelagos such as Hawaii are best of all because such islands, though themselves isolated, are often sufficiently close to insure occasional interchange of species. When a species of one island colonizes another, the two populations become increasingly different as time and evolution proceed. Additional colonizations will often follow and may result in the development of a species flock. Such flocks of related species as the Hawaiian honey creepers (birds) of the family Drepanididae or the giant land turtles of the Galapagos, provide interesting examples of the ways in which natural selection and random genetic changes can operate in small populations.

The huge Pacific plate, which underlies the greater part of the Pacific Ocean, is the largest tectonic plate in the world (Fig. 16). Its primary movement is from east to west so that along its western borders from the Aleutians to south of New Zealand, there occur a whole series of island arcs as the result of local subduction processes. The Pacific plate also contains many sites of intraplate volcanism, places where island chains are created above hot spots or mantle plumes. The result of

148

these structural activities is an enormous expanse of scattered islands and archipelagos that extend more than one-third of the way around the world.

In general, the islands of the Pacific plate demonstrate a biogeographic pattern that has three notable characteristics (Kay, 1979): (1) the flora and fauna all exhibit definite western Pacific relationships; (2) there is a west to east diversity gradient; and (3) there is sometimes an abrupt elimination of certain groups from west to east. It is evident that the third phenomenon has not occurred in a random manner for it is those groups that have the least dispersal ability over saltwater that drop out the quickest. The true (primary) freshwater fishes do not extend appreciably beyond Wallace's Line which marks the edge of the mainland – large island continental shelf. Terrestrial mammals, amphibians, and reptiles extend somewhat farther east. None of these vertebrates, with the exception of bats, extends appreciably into Polynesia (Cranbrook, 1981). In contrast, land birds are more widespread as are

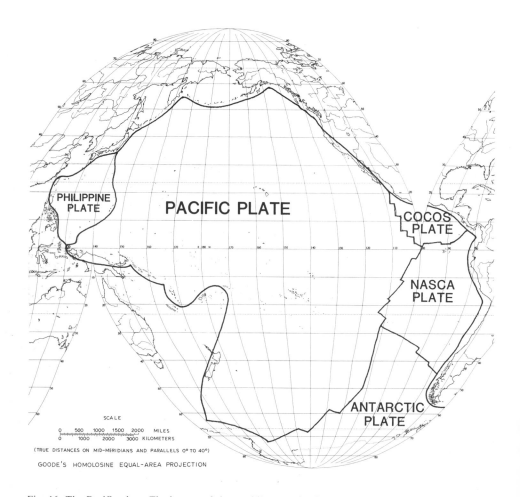

Fig. 16. The Pacific plate. The largest of the world's tectonic plates.

spiders, insects, and small species of land snails. Ferns and angiosperm plants in general are good colonizers and have achieved broad distributions in Polynesia.

The three notable characteristics of the Pacific plate biogeographic pattern apply also to the shallow-water marine biota. Among the most isolated island groups, the percentage of endemics is markedly lower than it is among the terrestrial organisms. In Hawaii for instance, 80–90% of the terrestrial snails are endemic but the endemism rate among the marine molluscs is only about 20% (Kay, 1979). In most tropical marine invertebrate groups, the great majority of species possess planktonic larvae that can be carried by ocean currents. The same is true for many shallow-water fish species. Thus, a place like the Hawaiian Archipelago, despite its isolation, is likely to receive some genetic reinforcement for its marine inhabitants while the terrestrial populations remain more deprived of such contact.

In 1982, Springer published an extensive study, primarily devoted to the shore fishes of the Indo-Pacific, which was intended to demonstrate that the Pacific plate defined "a biogeographic region of major significance". Previously, it had been observed that the most diverse marine faunal area in the world lies in the triangular area formed by the Philippines, the Malay Peninsula, and New Guinea (Briggs, 1974). But this area by itself could not be considered a distinct biogeographic province because the adjoining areas, although demonstrating notable reductions in species diversity, do not as a rule possess species that are not also found in the central triangle. Consequently, it was determined that the heart as well as the greater part of the Indo-West Pacific region was occupied by one huge province. The Indo-Polynesian province was considered to extend all the way from the entrance to the Persian Gulf to the Tuamotu Archipelago, and from Sandy Cape on the coast of eastern Australia to the Anami Islands in southern Japan. All of Polynesia was included with the exception of the Hawaiian Islands, the Marquesas, and Easter Island (Fig. 17).

The patterns of species diversity for almost all the shallow-water marine groups that are well represented in the Indo-Pacific indicate a peak in the East Indies (central triangle) area. From that area, the diversity tends to drop off in all directions. The phylogenetic relationships of certain groups that have been recently investigated indicate that the East Indies is inhabited by the most advanced species with the more primitive ones occupying peripheral ranges (Briggs, 1984a). It appears that the East Indies have been operating as a center of evolutionary radiation for the marine shore biota of the Indo-West Pacific region and to some extent, for the rest of the tropical marine world. The hermatypic corals provide a good example of this kind of pattern. In this group, the geological history of the various genera is well enough known so that their age spans have been estimated (Stehli and Wells, 1971). Thus we have direct evidence that the proposed center of origin in the East Indies is inhabited by the youngest genera and that the average generic age increases with distance from the center. As the age increases, the generic diversity decreases so that a reciprocal relationship between age and diversity is demonstrated. Stehli and Wells concluded that this kind of pattern suggested that genera evolve in the high-diversity region and, through time, extend their ranges to the peripheral regions of higher environmental stress.

Except for extremely isolated areas such as the Hawaiian Islands, the Marquesas,

150

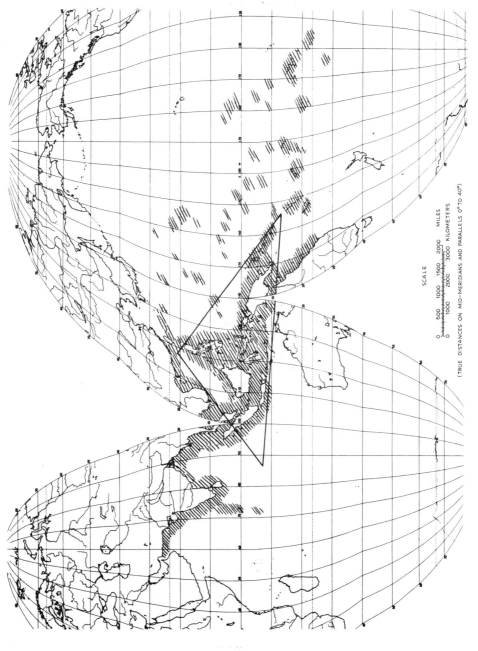

Fig. 17. The Indo-Polynesian province. The area of highest species diversity for most tropical marine animal groups is located within the central

SCALE

0   500   1000   1500   2000   MILES
0   1000   2000   3000   KILOMETERS

(TRUE DISTANCES ON MID-MERIDIANS AND PARALLELS 0° TO 40°)

and Easter Island, the great majority of species reported from Polynesian localities appear to be widespread in the Indo-West Pacific. It is this lack of local endemism that has resulted in the inclusion of most of the Pacific plate area in the Indo-Polynesian province. How much does a local biota have to differ from a parent one in order to merit recognition as a province? It has been suggested that if 10% or more of the species are endemic to a given area, it should be designated as a separate province (Briggs, 1974). The question raised by Springer's (1982) work is whether the Pacific plate, on the evidence provided by the distributional patterns of its shore fishes, should be recognized as a separate province. The facts, as gathered by Springer, are that 1312 shore fish species have been reported from the Pacific plate but only 48 of them or 6.30% are widespread endemics. Other endemics occur within the plate area but are confined to small parts or to single islands.

One can, of course, add up the endemics that are confined to such places as Hawaii or Easter Island, and those that are confined to groups of adjacent islands, combine them with the 48 general plate species, and reach a total endemism figure of about 20% (Springer, 1982). But this procedure is not justifiable in evolutionary biogeography where an important objective is the identification of the locality of speciation and the causes for its occurrence. As Scheltema and Williams (1983) have pointed out, in their article on the long-distance dispersal of planktonic molluscan larvae, the Hawaiian Archipelago and several other island systems have arisen as a consequence of plate motion over a mantle plume. Since the sea floor is much older than the origin of these island chains, no connection can ever have been possible with other land masses and it follows that colonization by marine species must, necessarily, have resulted from long distance dispersal.

A more recent work (Kay, 1984) dealt with patterns of speciation in the Indo-West Pacific among several groups of shallow-water marine animals. The author paid particular attention to the number of species that occurred on the Pacific plate as compared to the number that were confined to it. However, this analysis has a credibility problem. For example, 202 molluscan species are said to occur in the plate but we know that about 2000 species occur at the New Hebrides alone (Salvat, 1967); Kay gave a figure of 200 fish species for the plate when Springer (1982) had referred to 1312 species. The use of such figures gives a very high average endemism figure for the plate (40%) whereas, in reality, it is probably less than 10%.

The East Indies evolutionary center is located primarily on the continental shelf of the mainland and of the large islands that extend out to and including New Guinea. A number of the shore fish families do not extend out on the Pacific plate or occur only on its margin. This rather sharp drop in the eastward diversity of such families has been related to the presence of the plate boundary (Springer, 1982) but most of these families are ones that are best adapted to continental or large-island shorelines where there are estuaries and reduced salinity conditions. Many aspects of the ecology of the coral reef environment found in the waters of the scattered islands of Polynesia are very different from those of the bays and estuaries that lie to the west. It is most probable that ecological factors have prevented the eastward spread of many groups of the marine biota. Although the entire Pacific plate does not seem to comprise a distinct biogeographic province, important parts of it such as certain isolated islands and island groups, the oceanic basins, and the deep-sea trenches, certainly do.

152

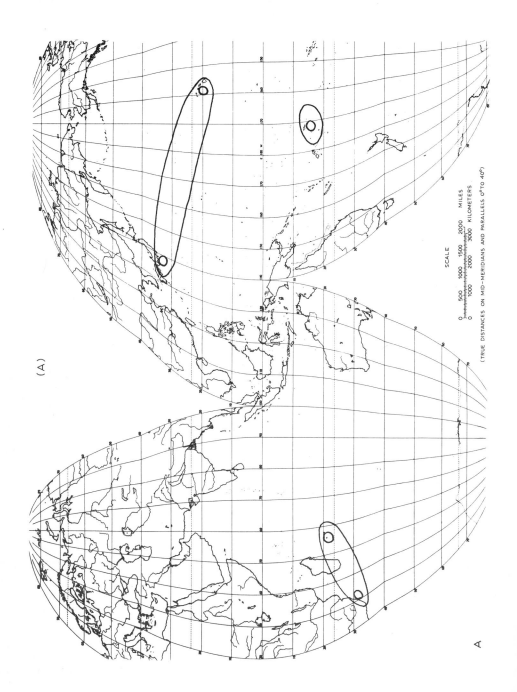

(A)

SCALE

| 0 | 500 | 1000 | 1500 | 2000 | MILES |

| 0 | 1000 | 2000 | 3000 | KILOMETERS |

(TRUE DISTANCES ON MID-MERIDIANS AND PARALLELS 0° TO 40°)

A

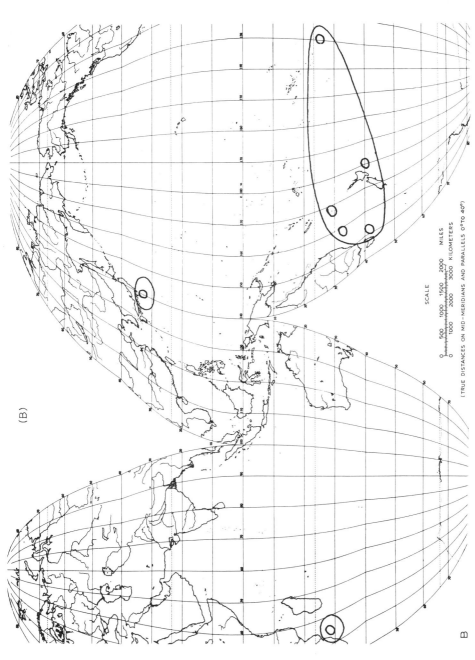

(B)

SCALE

0   500   1000   1500   2000   MILES

0   1000   2000   3000   KILOMETERS

(TRUE DISTANCES ON MID-MERIDIANS AND PARALLELS 0° TO 40°)

B

Fig. 18. Geographical distributions of two Indo-West Pacific marine fishes. A. *Myripristis chryseres*; B. *Pseudocaranx dentex*. Each have antitropical patterns in the western Pacific and relict populations in the western Indian Ocean.

In regard to the few shore fish and marine invertebrate species that appear to be broadly distributed on the Pacific plate but confined to it, there is another explanation for their presence aside from speciation on the plate. Many of them may be relatively old species that have become extinct in the central tropics and are being replaced by younger species from the East Indies evolutionary center. As the extinction and replacement process goes on, the distribution of the older species will be converted from a broad, continuous pattern to one in which it may occupy only the fringes of its original range (i.e., the outer parts of the Pacific plate). Such species are considered to be geographical and phylogenetic relicts.

A common type of fringe distribution has been identified as anti-tropical or anti-equatorial (Randall, 1982). This pattern usually pertains to species belonging to tropical genera that are not found within the central tropics but are relegated to its northern and southern edges. There are three explanations for such distributions: (1) transgression across the tropics was possible during a glacial period when the climate was much colder; (2) isothermic submersion by which species can extend across the tropics by swimming at greater depths where the temperature is cooler; and (3) the relict theory which envisions an extinction in the central tropics leaving relict populations in the cooler waters to the north and south.

The transgression theory, which was first proposed by Darwin (1859), was the most popular for many years until the history of sea surface temperature during the glacial periods became better known. It then became clear that the tropics had not became greatly altered during the Pleistocene. The isothermic submersion theory was, according to Ekman (1953), first proposed by Sir James Ross. For species that are capable of living in deep as well as shallow water, this submersion theory is reasonable. In fact, a number of antitropical species demonstrate continuous populations through the deep waters beneath the surface of the tropics (Briggs, 1974).

It was Theel (1885), in his work on the holothurians of the "Challenger" expedition, who proposed the relict theory to account for antitropical or "bipolar" distributions. As Randall (1982) has noted, there are in the Indo-West Pacific a number of examples of anti-tropical distribution. Such relicts often occur on the outermost (eastward) portions of the Pacific plate as well as to the north and south, but some are also represented by isolated populations in the western Indian Ocean (Fig. 18). These are just the kinds of patterns that one might expect to find as the result of extinctions followed by replacements from the evolutionary center in the East Indies. For the many species of shallow-water marine animals and plants that demonstrate antitropical distributions, and are not capable of living in deep water, the relict theory appears to be the best explanation.

SUMMARY

When oceanic plates diverge they form prominent, mid-ocean ridges that may extend thousands of meters above the surrounding sea floor. These elongate elevations tend to separate the various oceanic basins. When the plates converge, one is usually subducted beneath the other. The downward movement of the plate margin, as it is being subducted, results in the formation of a deep-sea trench. Trench formation is often accompanied by volcanic activity which results in the creation of an island arc. When island arcs are formed far from the mainland shore, back-arc basins may also be established. Transform faults, which serve to accommodate the differing lateral motions of diverging plates, may also have biogeographic importance.

Modern knowledge of sea-floor topography and the fauna of the ocean basins has revealed that many basins have very distinct animal assemblages with high rates of endemism. The fishes, most molluscs, isopods, ascidians, and sponges appear to be highly restricted in their horizontal distributions while the echinoderms, hydroids, and bivalve molluscs have more broadly distributed species. The animals inhabiting the deep-sea trenches have become highly modified. The average degree of endemism in a trench environment is 58%. The greater part of the species diversity found in the modern deep sea may be attributed to the barriers erected and the dispersal opportunities afforded by the gradual movement of the plates that comprise the crust of the sea floor.

Plate movements resulting in continental separations and joinings have had important effects in the marine environment. As the Atlantic Ocean opened up, the continental shelf fauna on each side become less related. Similar effects took place as Africa separated from the Caribbean and South America and as Australia departed from the vicinity of Antarctica. The most dramatic effects on the biota took place as continents closed with one another such as Africa with Asia and Alaska with Siberia.

Almost all isolated islands are volcanic in origin. Some are formed as the result of volcanic activity along the mid-ocean ridges. Such islands increase in age as they are carried away from the ridges by lateral plate movement. As the sea floor cools and subsides, many islands sink to become sea mounts or guyots. Eventually, all are consumed by the trenches. Many island chains are formed far from plate margins by the action of a plate moving over a hot spot or mantle plume. A third cause of island formation is by volcanic action that takes place in the vicinity of plate subduction. Such islands form curved chains called island arcs.

Isolated oceanic islands of known ages are of great interest to evolutionary biologists for they provide small laboratories where experiments, carried on by nature for thousands or millions of years, may be examined. Island archipelagos such as those of Hawaii or the Galapagos are of particular interest for they often possess flocks of closely related species that are found nowhere else.

The huge Pacific plate is the largest tectonic plate in the world. It possesses an enormous expanse of islands and archipelagos that extend more than one-third of the way around the world. The islands of the Pacific plate demonstrate a biogeographic pattern that has three notable characteristics: (1) the flora and fauna all exhibit definite western Pacific relationships; (2) there is a west to east diversity gradient; and (3) there is sometimes an abrupt elimination of major groups from west to east. Among the most isolated islands, the endemism in the terrestrial biota tends to be much greater than in the marine biota.

The part of the world with the most diverse marine fauna is the triangular area formed by the Philippines, the Malay Peninsula, and New Guinea. This East Indies triangle lies at the heart of the Indo-Polynesian province of the Indo-West Pacific biogeographic region. In many groups of the shelf biota, it appears that the more advanced species are located in the East Indies with the primitive ones occupying peripheral habitats. Consequently, it has been suggested that the high diversity area of the East Indies operates as a center of evolutionary radiation that gradually supplies advanced species to the rest of the Indo-West Pacific.

Except for extremely isolated areas such as the Hawaiian Islands, the Marquesas, and Easter Island, the great majority of species reported from the Pacific plate are widespread in the Indo-West Pacific. A work on the shore fishes of the plate was intended to show that it comprised a distinct biogeographic region but only about 6.30% of the species turned out to be general plate endemics. Other endemics are found in the most isolated island areas but, with few exceptions, there is no reason to believe that they were once more widely distributed. Most local endemics have probably originated from colonizations as the result of long-distance dispersal by ocean currents.

A number of fish families that are well established in the East Indies do not extend out on the Pacific plate or do so only to a very limited extent. Such families are mainly ones that appear to be well adapted to continental and large island shorelines and may have difficulty in adjusting to the coral reefs of the Polynesian area. In general terms, the Pacific plate has probably been gradually picking up its marine biota from the East Indies via long-distance dispersal for millions of years. Although a few species appear to be widespread and confined to the plate, many of them are probably old species that have been replaced by younger relatives in the other parts of their original range. Thus, they survive on the plate as phylogenetic and geographical relicts.

A number of Indo-West Pacific shallow-water animal and plant species demonstrate antitropical distribution patterns. In the past, it has been suggested that such patterns were the result of tropical transgression during glacial periods or of isothermic submergence. But the sea surface of the tropics did not cool down very much during the glaciations and relatively few shallow-water species can also survive in the deep sea. Many antitropical patterns can be explained as resulting from extinctions that apparently spread outward from the evolutionary center in the East Indies. It has been suggested that such extinctions were the results of competition from newly evolved species (Briggs, 1974) but one cannot rule out other causes such as predation, parasites, or disease.

CONCLUSIONS

> . . . one can live in a prefabricated world, smugly and without question, or one can indulge perhaps the greatest human excitement: that of observation to speculation to hypothesis. This is a creative process, probably the highest and most satisfactory we know.
>
> John Steinbeck, *Between Pacific Tides,* 1939.

While it is possible to demonstrate that the amalgamation and subsequent dispersal of the continents has had direct effects on the evolution and distribution of the earth's biota, there are also indirect effects that may have had important biological consequences. Fischer (1981, 1984) pointed out that active, worldwide plate movements are accompanied by increased volcanism which in turn provides an increase in the amount of carbon dioxide supplied to the atmosphere-hydrosphere system.

The divergence of tectonic plates in the ocean basins promotes the development of mid-oceanic ridges. The latter structures displace water from the basins onto the continents to form epicontinental seas. Consequently, the area of the land to which the atmosphere loses carbon dioxide by weathering, is decreased, and with it the rate at which carbon dioxide is removed from the air and returned to the lithosphere. The net effect of these two processes is that during times of increasing rates of plate movement, the level of carbon dioxide in the atmosphere-hydrosphere system must rise (Fischer, 1981, 1984).

Berner et al. (1983) employed a computer model to examine many aspects of the carbonate-silicate geochemical cycle and its ultimate effect on atmospheric carbon dioxide. This study was later extended (Lasaga et al., 1985) to include reactions involving organic carbon and sulphur. The results indicated that the carbon dioxide content of the atmosphere is highly sensitive to changes in sea-floor spreading rate and continental land area. The carbon dioxide level predicted for the Cretaceous was about 13 times present-day values.

A rise in the carbon dioxide content of the atmosphere during plate activity would be sufficient to cause retention of much of the heat now lost to space by radiation. This greenhouse effect would imply a rise in the mean average surface temperature of the globe of one to several degrees. In general, the tropics would not get much warmer but the excess calories would be converted to latent heat by means of evaporating water. This excess would then be transported as atmospheric moisture to higher latitudes warming the polar and temperate regions.

A world existing in a greenhouse state would be very different than our present sphere. Apparently, the Cretaceous was a time when greenhouse conditions existed. The extensive inland seas, the moist tropical climate, and the high atmospheric level of carbon dioxide produced a luxuriant growth of plant life. The broad geographic extent of the tropics, which covered most of the globe except for warm-temperate areas near the poles, allowed the development of high levels of species diversity which may have stimulated evolutionary progress. It was a critical time for the production of most of the families of our modern animals and plants. But, in the Cretaceous, there were no cold-temperate or cold environments and the modern biota of the higher latitudes had to evolve or acclimate during the succeeding 65 Ma. The burning of fossil fuels is again increasing the atmospheric carbon dioxide. How much more will be required to start another greenhouse effect?

It is the place of the southern continents and land masses (and ocean plateaus) in the general scheme of dispersal that is poorly known. Having examined the biological data for the southern hemisphere, the next step is to attempt to correlate this information with the geological data to take a new look at the sequence of events that have been so important to the evolution and distribution of life on earth.

When one compares the biota of the high-latitude versus the low-latitude parts of the world, it becomes at once apparent that the former usually contains the older and more primitive organisms. Furthermore, there is a difference between the northern and southern hemispheres for in the south we find preserved many ancient forms that no longer live in the far north. How can we account for this state of affairs? In most major groups (both animals and plants) for which there is adequate fossil material, we can find examples of genera or families that have evidently evolved in the tropics and subsequently become widespread in both hemispheres. The next step has often been an extinction in the tropics leaving a bipolar or antitropical distribution to the north and south. Finally, in many cases, there has occurred an extinction in the north leaving relict populations on one or more of the southern land masses.

The striking accumulation of primitive relicts in the terrestrial southern hemisphere, particularly in Australia and southern South America, has long fascinated biogeographers and evolutionary biologists. It is interesting to see that the accumulation process apparently started soon after the continents began to disperse. Molnar (1984) noted that a labrynthodont amphibian had persisted in Australia until the Lower Jurassic whereas, in the rest of the world, none had survived beyond the end of the Triassic. He also noted that the Jurassic reptilian fauna of Australia contained species that had already become extinct in other parts of the world. The Australian monotremes (the duckbill and spiny anteaters) are famous relicts that possibly evolved from Triassic docodont ancestors (Colbert, 1980). And there are (or were in some cases) in the southern hemisphere the ancient ratite birds including the cassowaries, emus, moas, kiwis, rheas, ostriches, and elephant birds. The survival of the Australian lungfish, *Neoceratodus*, which closely resembles its Triassic ancestors, is remarkable. Well known relicts in New Zealand are the rhyncocephalian, *Sphenodon*, and the frog *Leiopelma*.

There are large numbers of lesser known relicts that, taken as a whole, comprise a significant proportion of the southern hemisphere, high-latitude biota. Among the birds, both South America and Australia possess many primitive families. In regard to insects, Mackerras (1970) has pointed out that primitive components of most orders of insects from mayflies to beetles are shared between Australia and South America and to a lesser extent with New Zealand and Africa. Primitive spiders show the same general pattern (Main, 1981). So do freshwater mussels, snails, and crayfish (p. 71). The oldest of the living conifer families is the Araucariaceae. Although it was once widespread, the living species now have a disjunct distribution in the southern hemisphere (Florin, 1963). The Winteraceae is considered to be the most primitive angiosperm family (Smith, 1972). Of the seven genera, six are restricted primarily to the Australian – New Zealand area.

In the northern continents, at least in the terrestrial environment, the earlier extinction of primitive forms may be related to the fact that the geographical areas are larger enabling the species diversity to be greater and the interspecific competition to be more intense. In recent years, whenever a primitive group with a discontinuous distribution in the southern hemisphere is found, there has been a strong tendency to say, "aha, Gondwanaland". We must keep in mind that not all southern discontinuous distributions originated on the southern continents and that the various isolated parts of the southern hemisphere have served as excellent refuges for old genera and families that originally came from the tropics. It seems apparent that, during the Cretaceous and throughout the Cenozoic, the three tropical centers of evolutionary radiation – Oriental, Ethiopian, Neotropical – have had a continuous effect on the biota of the world (Briggs, 1984a).

In the Introduction, reference was made to the enormous literature on plate tectonics that has been published in the last 15 years. These contributions are almost entirely geophysical in origin. The first modern attempt to combine the geophysical data with biological information was that of Barron et al. (1981). However, the biological input to that work was minor and the conclusions were still based primarily on geophysical evidence. This, and other works based on geophysical data, reached important conclusions that are not supported by biological evidence. Examples are: (1) a continuous separation between North America and Europe beginning about 160 Ma; (2) the idea that the Greater Antilles once occupied the position of the present Central America; (3) a Lower Jurassic separation of Africa from Europe; (4) a broad attachment between South America and Africa in the early Cretaceous; (5) a protracted connection between Australia and Antarctica lasting until the Paleocene; and (6) an extended period of isolation for India which did not end until the Miocene.

In contrast to the geophysical conclusions, the biological data indicate that: (1) Europe and eastern North America were joined throughout the Mesozoic and did not become separated from one another until the early Eocene; (2) the Greater Antilles as subaerial islands were not rafted from Central America but emerged in situ; (3) Africa must have remained attached to Europe until at least the Upper Jurassic and possibly until the beginning of the Cretaceous; (4) South America and Africa were probably separated in the early Cretaceous; (5) Australia has probably been separated from Antarctica since the Triassic; and (6) it seems clear that a terrestrial connection between India and Africa took place in the late Cretaceous and that the docking of India with Asia was achieved in the early Eocene.

Because such differences do occur and are exceedingly important from an evolutionary and biogeographic standpoint, it seems that it would be helpful to produce a series of maps depicting the distribution of land and sea from the Triassic (200 Ma) to the present. The maps have been drawn for the purpose of illustrating events in the history of continental relationships that have special significance for biogeography. The events, and their more important biogeographical consequences, are reviewed in chronological order:

The story, as it is commonly presented, begins with Pangaea (Map 1), most often as it was interpreted by Dietz and Holden (1970). Supposedly, the first rift took place, in the late Triassic, between North Africa and North America beginning the formation of the modern North Atlantic Ocean. Sometimes this rift is depicted as a complete north-south separation, dividing Pangaea into two united continents, Laurasia to the north and Gondwanaland to the south (van Andel, 1979; Nelson and Platnick, 1984). However, evidence from the distribution of the tetrapod faunas (Cox, 1974; Charig, 1979) indicates that a north-south connection persisted until the middle or the late Jurassic (Map 2) and possibly until the early Cretaceous.

It seems clear that the separation of Africa and South America began at their southern tips in the early Jurassic. However, the two continents probably remained connected in the Gulf of Guinea region until the Upper Jurassic (Map 2) and perhaps until the beginning of the Cretaceous. Analyses of the marine sediments in the South Atlantic (van Andel et al., 1977) indicated the presence of isolated, narrow marine basins between the two continents at the beginning of the Cretaceous. The relationships of the freshwater animals appear to indicate the presence of a narrow (filter bridge) attachment in the Upper Jurassic which may have been broken by the beginning of the Cretaceous. The proposition that the two continents were separated throughout the Cretaceous appears to be consistent with the lack of relationship among the birds, mammals, and angiosperm plants — groups that evolved rapidly during the entire Period. It is unfortunate that many maps illustrating early Cretaceous continental relationships (i.e. Barron et al., 1981) give the impression that Africa and South America were still broadly contiguous.

Epicontinental seas, which provided the initial barrier between Africa and South America, also probably played an important role in the relationship of the other southern land masses. The biota of New Zealand, both fossil and recent, gives no indications of terrestrial connections to Gondwanaland. Australia may have been connected in the Triassic but probably not after that period. Yet, both of these land masses are consistently depicted as integral parts of Gondwanaland throughout the Mesozoic. In fact, Australia is usually shown with a remaining connection to Antarctica as late as the Paleocene (Smith and Briden, 1977; Barron et al., 1981; Crook, 1981) or the Eocene (Hallam, 1981; van Andel, 1985).

India is the most controversial part of the Gondwanaland puzzle. Some of the more recent pre-dispersal reconstructions (King, 1980; Kennett, 1982) depict India as lying adjacent to Australia and Antarctica but far distant from the East African coast and Madagascar. Yet, the Mesozoic fossil relationships of India including dinosaurs, ancient conifer genera, and marine ostracods, indicate a relationship to the African coast rather than Australia. India probably separated from Madagascar to start its northward journey in the early Cretaceous (Map 3). While much of the geophysical evidence seems to show that India did not become fused with Eurasia until the Miocene, the fossils of freshwater fishes indicate that this event took place in the early Eocene.

In the case of Madagascar, the paleocontinental maps of Smith and Briden (1977) showed a connection to Africa until the mid-Cretaceous but an earlier geological study of the Mozambique Channel (Flores, 1970) indicated that, during the Cretaceous, there was a continuous separation from the African mainland. The absence of freshwater ostariophysan fishes indicates that the separation probably first took place at least by the Upper Jurassic. This indication now has some reinforcement for Tarling (1980) observed that the retention of Madagascar in its present position for the Lower Jurassic is more geologically and geophysically realistic than conventional reconstructions. More recently, Rabinowitz et al. (1983) indicated that the separation of Madagascar began in the middle Jurassic. Considering the geological and biological data that point to India's early relationship to both Madagascar and East Africa, it seems reasonable to suggest that an initial split in the mid-Jurassic between Madagascar and Africa resulted in an eastward movement of Madagascar-India. The break between Madagascar and India probably occurred later (early Cretaceous).

By the early Cretaceous (Map 3), the terrestrial portions of all the southern continents are separated and a complete Tethys Sea extends through the equatorial region. Active plate movement has resulted in high sea levels. Portions of Africa and Australia are inundated and the Turgai Sea (which first formed in the mid-Jurassic) separates Euramerica from Asia. In the mid-Cretaceous (Map 4), a Mid-Continental Sea has separated Westamerica from Euramerica, creating three northern continents. The presence of the two northern, epicontinental seas is marked by the presence of distinct provinces of marine animals, especially the bivalve molluscs and ammonoids.

In the late Cretaceous (Map 5), two continental connections of significant importance took place. To the north, West America and Asia became attached by the Bering Land Bridge. This separated the marine biota of the Bering Sea – Arctic Ocean (a vicariant event) while creating a migration corridor (a dispersal opportunity) for the terrestrial and freshwater biota. Late Cretaceous migratory movements across Beringia were probably made by the ancient paddlefishes (Polyodontidae), an ancestral mooneye (Hyodontidae), and catfishes (Ictaluridae). Also possibly salamanders (Cryptobranchidae), lizards (varanoid necrosaurs, varanoids), and a few early mammals. The Bering connection reduced the number of northern continents from three to two. This configuration is borne out by the distribution patterns of dinosaurs (tyrannosaurs and protoceratopsids only in Asiamerica and iguanodonts and primitive ankylosaurs only in Euramerica) and by the recognition of separate floral provinces for the same continental areas.

To the south, there was evidently a late Cretaceous connection between India and northern Africa. In general, the Cretaceous/Paleocene fauna of India does not exhibit signs of an extended isolation but, instead, demonstrates a close relationship to Africa and Madagascar. An Upper Cretaceous dinosaur seems to be common to all three areas. There are also Cretaceous/Paleocene pelobatid frogs and anguid lizards. These are holarctic forms which could not have been present in Madagascar but could possibly have existed in northern Africa. There must have been a connection between India and northern Africa or else India might have somehow bridged the gap between Africa and Asia.

By the late Cretaceous, there was evidently a filter bridge (probably an archipelago) between North and South America. The recent discovery of Cretaceous marsupials in Peru has made it necessary to reconsider the predominant theory that marsupials arose in North America then migrated south. It is possible that both marsupials and the primitive condylarths arose in South America and then migrated north. Among the reptiles, the late Cretaceous hadrosaurs, common in the northern hemisphere, have now been recorded from several localities in the Upper Cretaceous of Argentina. And the snake genus *Coniophis,* of uncertain family status, is known from the late Cretaceous to the Eocene of North America and from the late Cretaceous of Bolivia.

By Paleocene times (Map 6), the epicontinental seas had receded to a considerable extent and the Mid-Continental Sea dried up, producing one huge northern continent, a Eurasiamerica. For most groups, there is much better fossil evidence that gives a more complete picture of traffic that took place across Beringia and between North and South America. In South America, the edentates began to blossom but a single fossil has shown up in the late Paleocene of Asia. In a like manner, the notoungulates began a major diversification in South America but a few species have shown up in the late Paleocene of Asia. Three lizard groups, the teiids, anguids, and skinks probably reached South America from the north. Other elements of the herpetofauna dispersed from North America into Central America.

In the early Paleocene, there was a sudden origin of mammalian families and genera, and by the late Paleocene, there appeared in North America, probably as the result of immigration from Asia across Beringia, adapid and omomyid primates, hyaenodontids, and possibly condylarths, perissodactyls, and artiodactyls. Three genera of the lizard family Scincidae probably also came from Asia. There is also evidence of the movement of some temperate angiosperm plants and insects across Beringia. At this time, Europe was still connected to eastern North America with the mammalian fossils indicating one homogeneous fauna.

The important biogeographic events in the Eocene (Map 7) are the separation of Europe from North America, the docking of the Indian continent against Asia, and fossil evidence of the presence of marsupials in Antarctica. In regard to the North Atlantic connection, changes in the mammalian fauna of North America compared to Europe indicate a break in the early Eocene. This coincides with the presence of marine fossils at Spitsbergen which apparently indicate the first opening between the Arctic Ocean and the North Atlantic. The distribution patterns of salamanders and certain groups of freshwater fishes appear to be consistent with this interpretation. While most of the geophysical reconstructions show India fusing with Asia in the Miocene, fossils of freshwater fishes show that this event took place in the early Eocene. Also, by the Middle Eocene, mammals of Asiatic origin were in India.

The prevailing theory that Australia received its original marsupial fauna from South America via Antarctica, has been reinforced by the find of late Eocene marsupial remains in Antarctica. Considering that the Antarctic specimens closely resemble taxa that lived in South America about 50 Ma, it has been suggested that the migration to Antarctica took place at about that time. Traffic across the Bering Land Bridge was considerable. Mammals such as the mesonychids, taperoids, hyracodontids, rhinocerotoids, and chalicotheres moved in one or both directions. A resurgence of similarity between the mammals of Asia and North America at the end of the Eocene and beginning of the Oligocene, has been attributed to the influence of the Bering Land Bridge. Several freshwater fish families, that are represented in the early and mid-Eocene Green River Formation of North America demonstrate Asiatic relationships. Major elements of the Asian Arcto-Tertiary flora arrived in North America.

There is also evidence of migratory activity between the Americas in approximately Eocene time, a primitive primate probably entered South America to begin the evolution of the New World monkeys. The caviomorph rodents probably also reached South America in the late Eocene or early Oligocene. South American plants became established in the tropical portions of the North American region from at least Eocene time onward. The freshwater fish family Poeciliidae and several families of the herpetofauna must have been established in Central America by the Eocene. These are old groups that demonstrate considerable evolutionary progress (producing distinct tribes and genera) in the Central American area.

The Oligocene began with a sea-level fall which resulted in the desiccation of the Turgai Sea and the joining of Europe to Asia. The establishment of this connection had a major impact on the mammalian fauna of Europe and undoubtedly on other groups of plants and animals. Several new mammal families that apparently originated in Asia arrived in North America via Beringia. Salamanders of the family Plethodontidae probably traveled the opposite direction to reach Asia and eventually Europe. The frog genus *Bufo* reached Central America from South America and later invaded North America. Two other frog families may have also reached North America from the south at about this time. During the Miocene (Map 8), there was an expansion of the Antarctic ice cap and the sea level dropped even more. Africa became attached to Eurasia via the Arabian Peninsula putting an end to the Tethys Sea. This was followed by an additional attachment in the Gibraltar area which isolated the Mediterranean Sea causing it to dry up (the Messinian Regression). In Africa, a dramatic faunal upheaval occurred during the late Oligocene/early Miocene when 29 new families and 79 new genera of mammals made their appearance. Archaic suids, canids, viverrids, and felids arrived from Eurasia. By the late Miocene, another 18 families appeared. Certain Eurasian freshwater fish families (Cyprinidae and probably the Cobitididae, Mastacemblidae, Channidae, Anabantidae, and Belontiidae) evidently were able to enter Africa at this time.

In the early Miocene, a great wave of mammalian immigration took place when some 16 genera came across Beringia and established themselves in North America. Also the presence of Miocene mammalian fossils of North American affinity in Panama indicates a fairly easy passage from North America to that point. A North American raccoon managed to reach South America. In the opposite direction came two genera representing two different families of ground sloths. In the West Indies, the uplift of the Greater Antilles had apparently taken place or was being completed by Miocene times. The phylogenetic relationships of the freshwater fishes and of various groups of terrestrial animals suggest that the larger islands were beginning to accumulate their faunas by means of oversea dispersal. By the mid-Miocene (Map 8), Australia had moved far enough north so that it began to pick up tropical organisms from the southeast Asian area via New Guinea.

In the Pliocene, traffic across the Bering Land Bridge continued until about 3.7 Ma. At this time, the sea level rose and the Bering Land Bridge was inundated. For the terrestrial flora and fauna this was an important vicariant event for it resulted in an immediate cessation of gene flow that lasted for more than a million years. For the marine fauna and flora a migration corridor, that had been closed since the late Cretaceous, was suddenly opened. During the earlier Tertiary, the cold-temperate marine biota of the North Pacific had evolved to a very high level of diversity. When the opportunity became available, many of the North Pacific species rapidly invaded the Arctic Ocean and the North Atlantic. This added greatly to the species diversity in the latter two oceans. At about 2.5 Ma the sea level fell once again and the Bering Land Bridge was restored.

In regard to the Caribbean connection, the final significant barrier between Central and South America, the Bolivar Seaway, gradually became narrower. Toward the end of the Pliocene, the first great group of South American mammals appeared in North America. These consisted of a sloth, three armadilloid genera, two hystricognath rodents, a porcupine, and a capybara. The Greater Antilles continued to pick more terrestrial species by fortuitous means. In the early Pliocene, placental rodents of the family Muridae reached Australia, probably by island hopping down the East Indian chain. The African mammalian fauna continued to be enriched by immigrants from Asia. Among them were the camel, the modern giraffe, the horse genus *Equus,* and a sabertoothed cat.

Important changes of biogeographic consequence that occurred in the Pleistocene to Recent epochs are: the completion of the Central American Land Bridge, the intermittent establishment of the Bering Land Bridge, and the extensive glaciations in the northern hemisphere. Toward the close of the Tertiary and the beginning of the Quaternary, about 3 Ma, the Bolivar Seaway became closed allowing the formation of a terrestrial corridor all the way to North America. Thus the "Great American Biotic Interchange" got underway. While the completion of the isthmian connection provided migration opportunities for the terrestrial and freshwater biota, it separated the tropical marine biota of the New World into eastern Pacific and western Atlantic components.

Both the creation of the terrestrial corridor (a dispersal event) and the division of the marine habitat (a vicariant event) had important evolutionary consequences. On land, representatives of 15 families of North American mammals entered South America and seven families migrated in the opposite direction. In South America, the effect was catastrophic and resulted in the extinction of the unique notoungulates, litopterns, and marsupial carnivores. When they reached South America, many of the northern mammalian groups underwent a rapid diversification. Much of the South American biota moved into the tropics of Central America. In the marine habitat, 3 million years of separation has resulted in the species on each side of the isthmus becoming almost completely distinct from one another.

During each of the major glaciations of the Pleistocene, the sea level dropped and the Bering Land Bridge was exposed. A phase of active intermigration took place beginning about 1.8 Ma. Eurasian forms reaching North America included jaguars, bovids, a mammoth, a caribou, and lemmings. In the late Pleistocene, other imports were the muskox, moose, bison, a weasel, and a fox. *Homo sapiens* probably arrived during the most recent (Wisconsin) ice age. During the interglacial stages, the Bering Land Bridge was flooded blocking the intercontinental terrestrial link but allowing the marine biota to pass between the Arctic Ocean and the Bering Sea.

In the Pleistocene, ice sheets advanced and retreated over North America and northern Eurasia. The ice-covered areas were almost completely uninhabitable but animal and plant distributions were also affected in other ways. Arctic species were driven south or else confined to refugia that remained ice-free. Associated climatic changes had far-reaching effects on the biota in many parts of the world and the melt water from the glaciers directly affected the distribution of aquatic life.

A general conclusion of some importance that affects the separation sequence as it is usually presented, is that Gondwanaland, in terms of an amalgamation of southern continents that was separate from a northern Laurasia, probably never existed in the Mesozoic. Toward the end of the Jurassic, when the separation of Africa from Euramerica took place to form a complete Tethys Sea, South America was in the process of a final separation from Africa. To the south, India – Madagascar had already separated from East Africa and Australia had probably not been in touch (in a terrestrial sense) with Antarctica since the Triassic. New Zealand was probably an island throughout the Mesozoic.

Plate tectonics obviously had very important effects on the evolution of higher forms of life. Certainly, the world would not be blessed with its current state of animal and plant diversity if the post-Triassic continental dispersal had not taken place. An important question is, are the tectonic plates constantly shifting at random or is there a long-term pattern to such movements? Is there a 300-Ma Phanerozoic supercycle as Fischer (1984) has suggested? Also, have extraterrestrial events, such a strikes by comets or asteroids, affected the movements of the plates? Do we have tectonic plates on earth because limestone-generating life evolved here (Anderson, 1984)? The answers to these questions must probably come from the fields of geophysics, astrophysics, and geochemistry but they are of great importance to historical biogeography.

Appendix

BIOGEOGRAPHER'S MAPS

These are biogeographer's maps intended to show the approximate configurations of land and sea for the indicated time period. The Lambert equal-area projections allow a minimum distortion of continental shapes and have the additional advantage of showing both poles on the same map.

MAP 1 – TRIASSIC

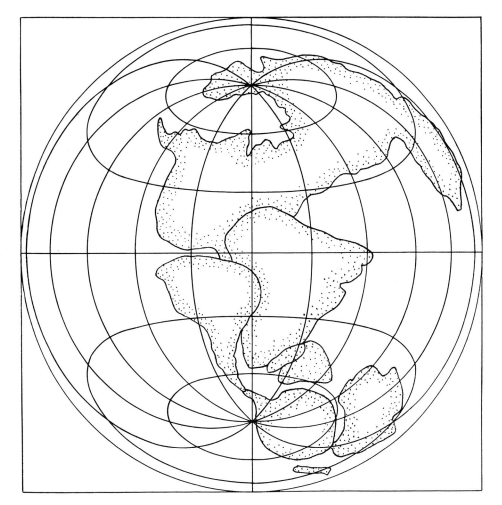

This map represents Pangaea, the global amalgamation of continents. The northern portion is sometimes referred to as Laurasia and the southern part as Gondwana or Gondwanaland.

MAP 2 – LATE JURASSIC

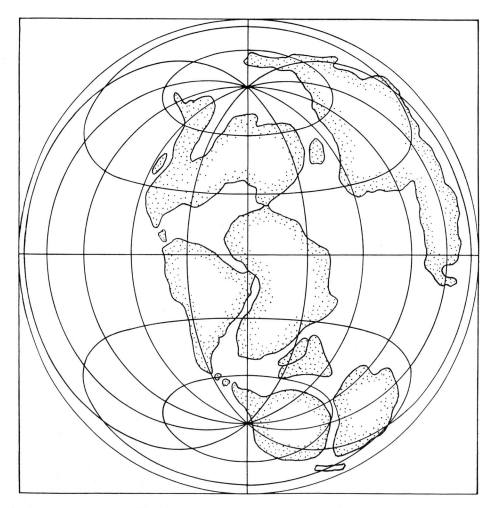

After some 60 Ma, considerable movement has taken place. Epicontinental seas have separated Australia – New Guinea from Antarctica and Europe from Asia. Madagascar – India has broken away from Africa, Euramerica still has a narrow connection to Africa and South America a narrow attachment to Africa. The Atlantic Ocean has developed extensively.

MAP 3 – EARLY CRETACEOUS

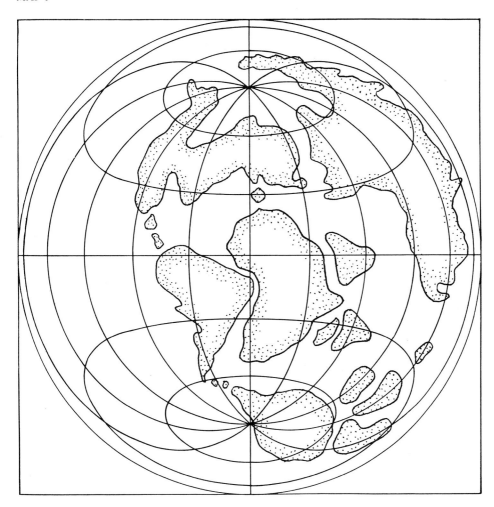

The northern continents are now completely separated from the southern ones. South America is detached from Africa. Epicontinental seas have grown splitting off northeastern Africa and dividing Australia – New Guinea into three parts. The circumtropical Tethys Sea is evident. India has separated from Madagascar.

MAP 4 – MID-CRETACEOUS

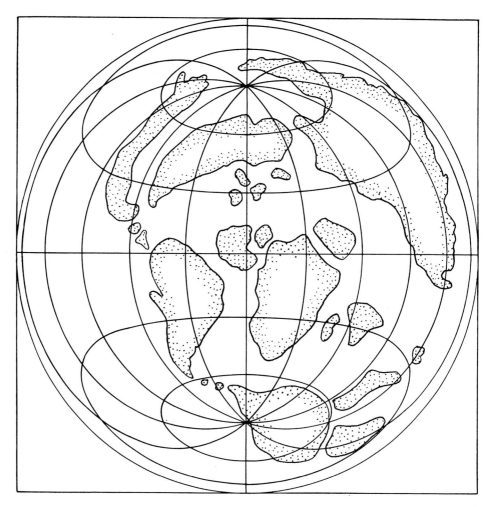

Further development of epicontinental seas creates three terrestrial continents in the north, Westamerica, Euramerica, and Asia. Southern Europe is a series of islands, northern Africa is fragmented, and Australia – New Guinea exists in two widely separated parts.

MAP 5 – LATE CRETACEOUS

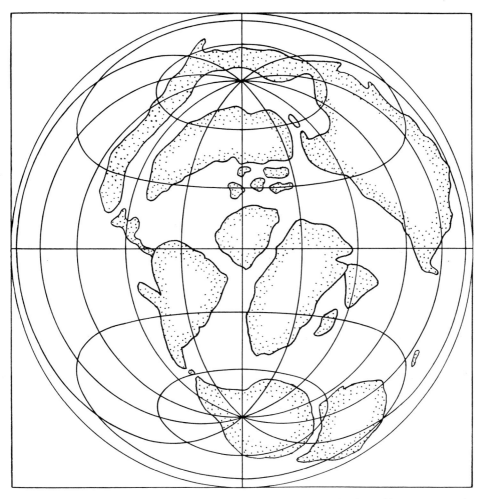

Westamerica becomes joined to Asia across the Bering Strait. India, during its rapid movement north-ward, has become attached to North Africa. Epicontinental seas are still high.

MAP 6 – PALEOCENE

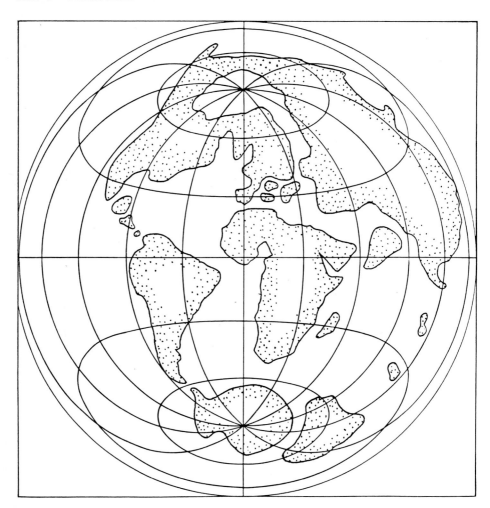

Epicontinental seas are receding. There is now only one northern continent. Southern Europe and parts of Africa are still inundated. India is approaching Asia. Portions of Central America are more fully emerged.

MAP 7 – EOCENE

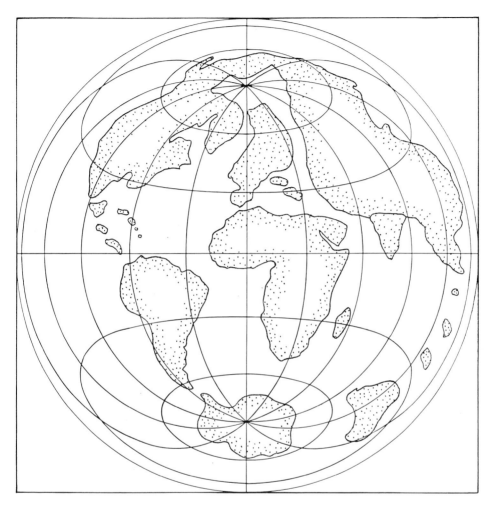

Europe and eastern North America have separated between Greenland and the Scandinavian Peninsula. India has collided with Asia producing the Himalayan uplift. Australia – New Guinea is now rapidly moving northward.

MAP 8 – MIOCENE

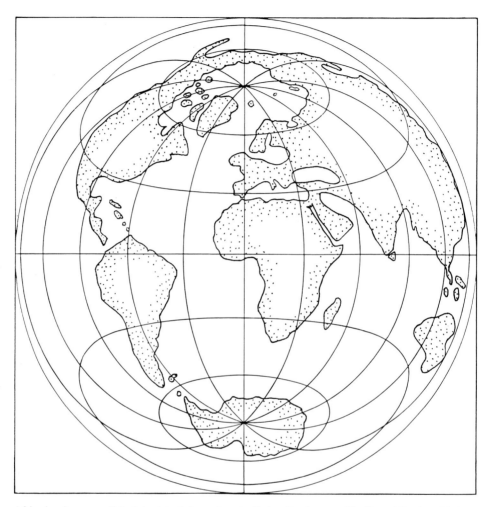

Africa has become solidly joined to Asia so that the Tethys Sea is gone. The Turgai Sea has dried up permitting Europe to become attached to Asia. Both the North and South Atlantic have become large oceans. The Bering Land Bridge is still evident, South America is still isolated, and Australia – New Guinea is approaching its present position.

MAP 9 – PRESENT DAY

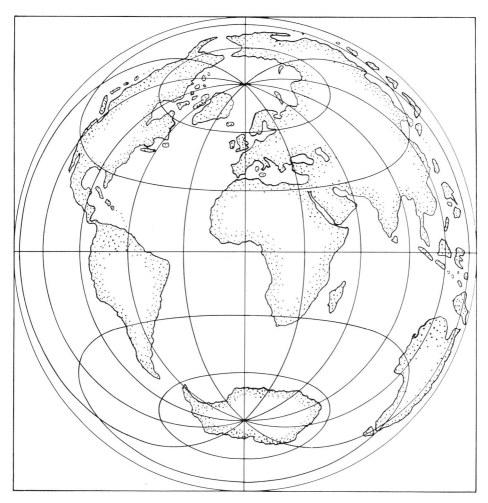

The Bering Land Bridge has become flooded, the Central American isthmus has been established, and the Black and Caspian Seas have been fully formed.

REFERENCES

Albignac, R., 1972. The Carnivora of Madagascar. In: *Biogeography and Ecology in Madagascar* (R. Battistini and G. Richard-Vindard, eds.). W. Junk, The Hague, pp. 667–682.

Anderson, D.L., 1984. The earth as a planet: paradigms and paradoxes. *Science*, 223: 347–355.

Anderson, T.H. and Schmidt, V.A., 1983. The evolution of Middle America and the Gulf of Mexico–Caribbean Sea region during Mesozoic time. *Geol. Soc. Am. Bull.*, 94: 941–966.

Araujo, R.L., 1970. Termites of the Neotropical Region. In: *Biology of Termites, Vol. II* (K. Krishna and F.M. Weesner, eds.). Academic Press, New York, N.Y., pp. 527–576.

Archer, M., 1984. Origins and early radiations of marsupials. In: *Vertebrate Zoogeography and Evolution in Australasia* (M. Archer and G. Clayton, eds.). Hesperian Press, Carlisle, W. Austr., pp. 585–625.

Archer, M., Flannery, T.F., Ritchie, A. and Molnar R.E., 1985. First Mesozoic mammal from Australia – an early Cretaceous monotreme. *Nature*, 318: 363–366.

Archibald, J.D. and Clemens, W.A., 1984. Mammal evolution near the Cretaceous-Tertiary boundary. In: *Catastrophes and Earth History* (W.A. Berggren and J.A. Van Couvering, eds.). Princeton University Press, Princeton, N.J., pp. 339–371.

Arden, D.D., 1975. Geology of Jamaica and the Nicaragua Rise. In: *The Ocean Basins and Margins: The Gulf of Mexico and the Caribbean* (A.E.M. Nairn and F.G. Stehli, eds.). Plenum Press, New York, N.Y., pp. 617–662.

Atkins, M.D., 1963. The Cupedidae of the world. *Can. Entomol.*, 95: 140–162.

Audley-Charles, M.G., 1981. Geological history of the region of Wallace's Line. In: *Wallace's Line and Plate Tectonics* (T.C. Whitmore, ed.). Clarendon Press, Oxford, pp. 24–35.

Audley-Charles, M.G., Hurley, A.M. and Smith, A.G., 1981. Continental movements in the Mesozoic and Cenozoic. In: *Wallace's Line and Plate Tectonics* (T.C. Whitmore, ed.). Clarendon Press, Oxford, pp. 9–23.

Axelrod, D.I., 1983. Biogeography of oaks in the Arcto-Tertiary Province. *Ann. Mo. Bot. Gard.*, 70: 629–657.

Axelrod, D.I., 1984. An interpretation of Cretaceous and Tertiary biota in polar regions. *Palaeogeogr., Palaeoclimatol., Palaeoecol.*, 45: 105–147.

Axelrod, D.I. and Raven, P.H., 1978. Late Cretaceous and Tertiary vegetation history of Africa. In: *Biogeography and Ecology of Southern Africa* (M.J.A. Werger, ed.). W. Junk, The Hague, pp. 77–130.

Baez, A.M. and de Gasparini, Z.B., 1979. The South American herpetofauna: an evaluation of the fossil record. In: *The South American Herpetofauna: Its Origin, Evolution, and Dispersal* (W.E. Duellman, ed.). *Mus. Nat. Hist. Univ. Kansas, Monogr.*, 7: 29–54.

Barbour, T., 1916. Some remarks upon Matthew's "Climate and Evolution". *Ann. N.Y. Acad. Sci.*, 27: 1–15.

Barron, E.J., Harrison, C.G.A., Sloan, J.L. and W.W. Hay, 1981. Paleogeography, 180 million years ago to the present. *Eclogae Geol. Helv.*, 74: 443–470.

Belyaev, G.M., 1966. *Hadal Bottom Fauna of the World Ocean* (English translation by Israel Program for Scientific Translations, Jerusalem, 1972).

Ben-Avraham, Z., 1981. The movement of continents. *Am. Sci.*, 69: 291–299.

Bennett, D.K., 1980. Stripes do not make a zebra, I. A cladistic analysis of *Equus*. *Syst. Zool.*, 29: 272–287.

Benson, R.H., 1984. The Phanerozoic "crisis" as viewed from the Miocene. In: *Catastrophes and Earth History* (W.A. Berggren and J.A. van Couvering, eds.). Princeton University Press, Princeton, N.J., pp. 437–446.

Berggren, N.A. and Hollister, C.D., 1977. Plate tectonics and paleocirculation. *Tectonophysics*, 38: 11–48.

178

Berner, R.A., Lasaga, A.C. and Garrels, R.M., 1983. The carbonate-silicate geochemical cycle and its effect on atmospheric carbon dioxide over the past 100 million years. *Am. J. Sci.,* 283: 641 – 683.

Berra, T.M., 1981. *An Atlas of Distribution of the Freshwater Fish Families of the World.* University of Nebraska Press, Lincoln, Nebr.

Besch, W., 1968. South American Arachnida. In: *Biogeography and Ecology in South America* (E.J. Fittkau et al., eds.). W. Junk, The Hague, pp. 723 – 740.

Bigalke, R.C., 1978. Mammals. In: *Biogeography and Ecology of Southern Africa* (M.J.A. Werger, ed.). W. Junk, The Hague, pp. 981 – 1098.

Biju-Duval, B., Bizon, G., Mascle, A. and Muller, C., 1982. Active margin processes: field observations in southern Hispaniola. In: *Studies in Continental Margin Geology* (J.S. Watkins and C.L. Drake, eds.). *Am. Assoc. Pet. Geol. Mem.,* 34: 325 – 344.

Bishop, J.A., 1967. The zoogeography of the Australian freshwater Decapod Crustacea. In: *Australian Island Waters and Their Faunas* (A.H. Weatherley, ed.). Australian National University Press, Canberra, A.C.T., pp. 107 – 122.

Blair, W.F., 1972. Summary. In: *Evolution in the genus Bufo* (W.E. Blair, ed.). University of Texas Press, Austin, Texas, pp. 329 – 343.

Blanc, C.P., 1972. Les reptiles de Madagascar et des Iles Voisines. In: *Biogeography and Ecology in Madagascar* (R. Battistini and G. Richard-Vindard, eds.). W. Junk, The Hague, pp. 501 – 614.

Boreske, J.R., 1974. A review of the North American fossil amiid fishes. *Bull. Mus. Comp. Zool. Harv.,* 146: 1 – 87.

Boufford, D.E. and Spongberg, S.A., 1983. Eastern Asian – eastern North American phytogeographical relationships – a history from the time of Linnaeus to the twentieth century. *Ann. Mo. Bot. Gard.,* 70: 423 – 439.

Bouillion, A., 1970. Termites of the Ethiopian Region. In: *Biology of Termites, Vol. II* (K. Krishna and F.M. Weesner, eds.). Academic Press, New York, N.Y., pp. 153 – 280.

Braithwaite, C.J.R., 1984. Geology of the Seychelles. In: *Biogeography and Ecology of the Seychelles Islands* (D.R. Stoddart, ed.). W. Junk, The Hague, pp. 17 – 38.

Brattstrom, B.H., 1964. Evolution of the pit vipers. *Trans. San Diego Soc. Nat. Hist.,* 13: 185 – 268.

Brenan, J.P.M., 1978. Some aspects of the phytogeography of tropical Africa. *Ann. Mo. Bot. Gard.,* 65: 437 – 478.

Brenner, G.J., 1976. Middle Cretaceous floral provinces and early migrations of angiosperms. In: *Origin and Early Evolution of Angiosperms* (C.B. Beck, ed.). Columbia University Press, New York, N.Y., pp. 23 – 44.

Briggs, J.C., 1966. Oceanic islands, endemism and marine paleotemperatures. *Syst. Zool.,* 15: 153 – 163.

Briggs, J.C., 1970. A faunal history of the North Atlantic Ocean. *Syst. Zool.,* 19: 19 – 34.

Briggs, J.C., 1974. *Marine Zoogeography.* McGraw-Hill, New York, N.Y.

Briggs, J.C., 1979. Ostariophysan zoogeography: an alternative hypothesis. *Copeia,* 1979: 111 – 118.

Briggs, J.C., 1984a. *Centers of Origin in Biogeography. Biogeographical Monographs, No. 1.* University of Leeds, Leeds.

Briggs, J.C., 1984b. Freshwater fishes and biogeography of Central America and the Antilles. *Syst. Zool.,* 3: 428 – 435.

Briggs, J.C., 1986. Introduction. In: *Zoogeography of North American Freshwater Fishes* (C.H. Hocutt and E.O. Wiley, eds.). John Wiley, New York, N.Y., pp. 1 – 16.

Brinkhurst, R.O. and Jamieson, B.G.M., 1971. *Aquatic Oligochaeta of the World.* University of Toronto Press, Toronto, Ont.

Brown, J.H. and Gibson, A.C., 1983. *Biogeography.* Mosby, St. Louis, Mo.

Brown, T.M. and Simons, E.L., 1984. First record of marsupials (Metatheria: Polyprotodonta) from the Oligocene in Africa. *Nature,* 308: 447 – 449.

Brown, W.L., Jr., 1973. A comparison of the Hylean and Congo-West African rain forest ant faunas. In: *Tropical Forest Ecosystems in Africa and South America: A Comparative Review* (B.L. Meggers, E.S. Ayensu and W.D. Duckworth, eds.). Smithsonian Press, Washington, D.C., pp. 161 – 185.

Browne, J., 1983. *The Secular Ark.* Yale University Press, New Haven, Conn.

Brundin, L., 1966. Transantarctic relationships and their significance, as evidenced by chironomid midges. *K. Sven. Vetenskapsvakad. Handl.,* 2: 1 – 472.

Bruun, A.F., 1957. Deep-sea and abyssal depths. In: *Treatise on Marine Ecology and Paleoecology, Vol. 1* (J.W. Hedgpeth, ed.). *Geol. Soc. Am., Mem.,* 67: 641 – 672.

Bull, P.C. and Whitaker, A.H., 1975. The amphibians, reptiles, birds and mammals. In: *Biogeography and Ecology in New Zealand* (G. Duschel, ed.). W. Junk, The Hague, pp. 231 – 276.

Burke, K., Cooper, C., Dewey, J.F., Mann, P. and Pindell, J.L., 1984. Caribbean tectonics and relative plate motions. In: *The Caribbean South American Plate Boundary and Regional Tectonics* (W.E. Bonini, R.B. Hargraves and R. Shagam, eds.). *Geol. Soc. Am., Mem.,* 162: 31 – 63.

Bussing, W.A., 1976. Geographic distribution of the San Juan ichthyofauna of Central America with remarks on its origin and ecology. In: *Investigations of the Ichthyofauna of Nicaraguan Lakes* (T.B. Thorson, ed.). School of Life Sciences, University of Nebraska, Lincoln, Nebr., pp. 157 – 175.

Bussing, W.A., 1985. Patterns of distribution of the Central American ichthyofauna. In: *The Great American Interchange* (F.G. Stehli and S.D. Webb, eds.). Plenum Press, New York, N.Y., pp. 453 – 473.

Cadle, J.E., 1985. The Neotropical colubrid snake fauna (Serpentes: Colubridae): lineage components and biogeography. *Syst. Zool.,* 34: 1 – 18.

Calabrese, D.M., 1980. Zoogeography and cladistic analysis of the Gerridae (Hemiptera: Heteroptera). *Misc. Publ. Entomol. Soc. Am.* 11: 1 – 119.

Calaby, J.H. and Gay, F.J., 1959. Aspects of the distribution and ecology of Australian termites. In: *Biogeography and Ecology in Australia* (A. Keast, R.L. Croker and C.S. Christian, eds.). W. Junk, The Hague, pp. 211 – 223.

Cande, S.C. and Mutter, J.C., 1982. A revised identification of the oldest sea-floor spreading anomalies between Australia and Antarctica. *Earth Planet. Sci. Lett.,* 58: 151 – 160.

Carlquist, S., 1965. *Island Life.* Natural History Press, Garden City, N.Y.

Carlquist, S., 1974. *Island Biology.* Columbia University Press, New York, N.Y.

Carpenter, A., 1977. Zoogeography of the New Zealand freshwater Decapoda: a review. *Tuatara,* 23: 41 – 48.

Charig, A., 1979. A new look at the dinosaurs. Mayflower Books, New York, N.Y.

Cheng, Z., 1983. A comparative study of the vegetation in Hubei Province. China, and in the Carolinas of the United States. *Ann. Mo. Bot. Gard.,* 70: 571 – 575.

Clarke, A.H., Jr., 1962. Annotated list and bibliography of the abyssal marine molluscs of the world. *Bull. Nat. Mus. Can.,* 181: 1 – 114.

Climo, F.M., 1975. The land snail fauna. In: *Biogeography and Ecology of New Zealand* (G. Kuschel, ed.). W. Junk, The Hague, pp. 459 – 492.

Cogger, H.G. and Heatwole, H., 1981. The Australian reptiles: origins, biogeography, distribution patterns and island evolution. In: *Ecological Biogeography of Australia* (A. Keast, ed.). W. Junk, The Hague, pp. 1333 – 1373.

Colbert, E.H., 1973. *Wandering Lands and Animals.* E.P. Sutton, New York, N.Y.

Colbert, E.H., 1980. *Evolution of the Vertebrates.* John Wiley and Sons, New York, N.Y., 3rd ed.

Colbert, E.H., 1982. Mesozoic vertebrates of Antarctica. In: *Antarctic Geoscience* (C. Craddock, ed.). University of Wisconsin Press, Madison, Wisc., pp. 619 – 627.

Collette, B.B. and Banarescu, P., 1977. Systematics and zoogeography of the fishes of the family Percidae. *J. Fish. Res. Board Can.,* 34: 1450 – 1463.

Cowley, D.R., 1978. Studies on the larvae of New Zealand Trichoptera. *N.Z. J. Zool.,* 5: 6391 – 6750.

Cox, C.B., 1974. Vertebrate paleodistributional patterns and continental drift. *J. Biogeogr.,* 1: 75 – 94.

Cracraft, J., 1973. Continental drift, paleoclimatology, and the evolution and biogeography of birds. *J. Zool. London,* 169: 455 – 545.

Cracraft, J., 1975. Historical biogeography and earth history: perspectives for a future synthesis. *Ann. Mo. Bot. Gard.,* 62: 227 – 250.

Cracraft, J., 1980. Moas and the Maori. *Nat. Hist.,* 89: 28 – 36.

Cracraft, J., 1983. Cladistic analysis and vicariance biogeography. *Am. Sci.,* 71: 273 – 281.

Cranbrook, E., 1981. The vertebrate faunas. In: *Wallace's Line and Plate Tectonics* (T.C. Whitmore, ed.). Clarendon Press, Oxford, pp. 57 – 69.

Craw, R.C., 1985. Classic problems of southern hemisphere biogeography re-examined. *Sonderdruck Z. Zool. Syst. Evolutionsforsch.,* 23: 1 – 10.

Crawford, A.R., 1974. A greater Gondwanaland. *Science,* 184: 1179 – 1181.

Croizat, L., Nelson, G. and Rosen, D.E., 1974. Centers of origin and related concepts. *Syst. Zool.,* 23: 265 – 287.

180

Cronin, J.E. and Sarich, V.M., 1980. Tupaiid and Archonta phylogeny: the macromolecular evidence. In: *Comparative Biology and Evolutionary Relationships of Tree Shrews* (W.P. Luckett, ed.). Plenum Press, New York, N.Y., pp. 293–312.

Crook, K.A.W., 1981. The break-up of the Australian Antarctic segment of Gondwanaland. In: *Ecological Biogeography of Australia* (A. Keast, ed.). W. Junk, The Hague, pp. 3–14.

Curray, J.R., Emmel, F.J., Moore, D.G. and Raitt, R.W., 1981. Structure, tectonics and geological history of the northeastern Indian Ocean. In: *Ocean Basins and Margins, Vol. 6, The Indian Ocean.* (A.E.M. Nairn and F.G. Stehli, eds.). Plenum Press, New York, N.Y.

Dana, T.F., 1975. Development of contemporary Eastern Pacific coral reefs. *Mar. Biol.,* 33: 355–374.

Darlington, P.J., Jr., 1938. The origin of the fauna of the Greater Antilles, with discussion of dispersal of animals over water and through the air. *Q. Rev. Biol.,* 13: 247–300.

Darlington, P.J., Jr., 1957. *Zoogeography: The Geographical Distribution of Animals.* John Wiley, New York, N.Y.

Darlington, P.J., Jr., 1959. Area, climate, and evolution. *Evolution,* 13: 488–510.

Darlington, P.J., Jr., 1965. *Biogeography of the Southern End of the World.* Harvard University Press, Cambridge, Mass.

Darwin, C., 1859. *On the Origin of Species by Means of Natural Selection.* John Murray, London.

Davis, G.M., 1979. The origin and evolution of the gastropod family Pomatiopsidae with emphasis on the Mekong River Triculina. *Acad. Nat. Sci. Philadelphia, Monogr.,* 20: 1–120.

de Beaufort, L.F., 1951. *Zoogeography of the Land and Inland Waters.* Sidgwick and Jackson, London.

Delevoryas, T., 1973. Postdrifting Mesozoic floral evolution. In: *Tropical Forest Ecosystems in Africa and South America: A Comparative Review* (B.J. Meggers, E.S. Ayensu and W.D. Duckworth, eds.). Smithsonian Press, Washington, D.C., pp. 9–19.

Dettmann, M.E., 1981. The Cretaceous Flora. In: *Ecological Biogeography of Australia* (A. Keast, ed.). W. Junk, The Hague, pp. 357–375.

De-Yuan, H., 1983. The distribution of Scrophulariaceae in the Holarctic with special reference to the floristic relationships between eastern Asia and eastern North America. *Ann. Mo. Bot. Gard.,* 70: 701–712.

Dietz, R.S. and Holden, J.C., 1970. Reconstruction of Pangaea: breakup and dispersion of continents. *Sci. Am.,* 223: 30–41.

Ding, S., 1979. A new edentate from the Paleocene of Guangdong. *Vertebr. PalAsiatica,* 17: 57–64.

Donnelly, T.W., 1985. Mesozoic and Cenozoic plate evolution of the Caribbean region. In: *The Great American Biotic Interchange* (F.G. Stehli and S.D. Webb, eds.). Plenum Press, New York, N.Y., pp. 89–121.

Dorst, J., 1972. The evolution and affinities of the birds of Madagascar. In: *Biogeography and Ecology in Madagascar* (R. Battistini and G. Richard-Vindard, eds.). W. Junk, The Hague, pp. 615–627.

Dott, R.H., Jr. and Batten R.L., 1971. *Evolution of the Earth.* McGraw-Hill, New York, N.Y.

Douglas, J.G. and Williams, G.W., 1982. Southern polar forests: the early Cretaceous floras of Victoria and their paleoclimatic significance. *Palaeogeogr., Palaeoclimatol., Palaeoecol.,* 39: 171–185.

Dransfield, J., 1981. Palms and Wallace's Line. In: *Wallace's Line and Plate Tectonics* (T.C. Whitmore, ed.). Clarendon Press, Oxford, pp. 43–56.

Duellman, W.E., 1979. The South American herpetofauna: a panoramic view. In: *The South American Herpetofauna: Its Origin, Evolution, and Dispersal* (W.E. Duellman, ed.). *Mus. Nat. Hist., Univ. Kansas, Monogr.,* 7: 1–28.

Duncan, R.A. and Hargraves, R.B., 1984. Plate tectonic evolution of the Caribbean region in the mantle reference frame. In: *The Caribbean – South American Plate Boundary and Regional Tectonics* (W.E. Bonini, R.B. Hargraves and R. Shagam, eds.). *Geol. Soc. Am., Mem.,* 162: 81–93.

Durham, J.W., 1985. Movement of the Caribbean plate and its importance for biogeography in the Caribbean. *Geology,* 13: 123–125.

Durham, J.W. and McNeil, F.S., 1967. Cenozoic migrations of marine invertebrates through the Bering Strait region. In: *The Bering Land Bridge* (D.M. Hopkins, ed.). Stanford University Press, Stanford, Calif., pp. 326–349.

Edmunds, G.F., Jr., 1972. Biogeography and evolution of Ephemeroptera. *Ann. Rev. Entomol.,* 17: 21–42.

Edmunds, G.F., Jr., 1975. Phylogenetic biogeography of mayflies. *Ann. Mo. Bot. Gard.*, 62: 251 – 263.

Edmunds, G.F., Jr., 1981. Discussion. In: *Vicariance Biogeography: A Critique* (G. Nelson and D.E. Rosen, eds.). Columbia University Press, New York, N.Y., pp. 287 – 297.

Edmunds, G.F., Jr., 1982. Historical and life history factors in the biogeography of mayflies. *Am. Zool.*, 22: 371 – 374.

Ekman, S., 1953. *Zoogeography of the Sea.* Sidgwick and Jackson, London.

Emerson, A.E., 1955. Geographical origins and dispersions of termite genera. *Fieldiana, Zool.*, 37: 465 – 521.

Endrody-Younga, S., 1978. Coleoptera. In: *Biogeography and Ecology of Southern Africa* (M.J.A. Werger, ed.). W. Junk, The Hague, pp. 797 – 821.

Erwin, T.L., 1979. The American connection, past and present, as a model blending dispersal and vicariance in the study of biogeography. In: *Carabid Beetles: Their Evolution, Natural History, and Classification* (T.L. Erwin, G.E. Ball, D.R. Whitehead and A.L. Halpern, eds.). W. Junk, The Hague, pp. 355 – 367.

Erwin, T.L., 1981. Taxon pulses, vicariance, and dispersal: an evolutionary synthesis illustrated by carabid beetles. In: *Vicariance Biogeography: A Critique* (G. Nelson and D.E. Rosen, eds.). Columbia University Press, New York, N.Y., pp. 159 – 183.

Estes, R., 1975. Fossil *Xenopus* from the Paleocene of South America and the zoogeography of pipid frogs. *Herpetologica,* 31: 263 – 278.

Estes, R., 1983. The fossil record and early distribution of lizards. In: *Advances in Herpetology and Evolutionary Biology* (A.G.J. Rhodin and K. Miyata, eds.). Museum of Comparative Zoology, Harvard University, Cambridge, Mass., pp. 365 – 398.

Estes, R. and Baez, A., 1985. Herpetofaunas of North and South America during the late Cretaceous and Cenozoic: evidence of interchange? In: *The Great American Interchange* (F.G. Stehli and S.D. Webb, eds.). Plenum Press, New York, N.Y., pp. 139 – 197.

Estes, R. and Reig, O.A., 1973. The early fossil record of frogs. In: *Evolutionary Biology of the Anurans* (J.E. Vail, ed.). University of Missouri Press, Columbia, Mo., pp. 11 – 63.

Estes, R. and Wake, M.H., 1972. The first fossil record of caecilian amphibians. *Nature,* 239: 228 – 231.

Evans, J.W., 1959. The zoogeography of some Australian insects. In: *Biogeography and Ecology in Australia* (A. Keast, R.L. Crocker and C.S. Christian, eds.). W. Junk, The Hague, pp. 150 – 163.

Fallow, W.C., 1979. Trans-North Atlantic similarity among Mesozoic and Cenozoic invertebrates correlated with widening of the ocean basin. *Geology,* 7: 389 – 400.

Ferris, V.R., Goseco, C.G. and Ferris, J.M., 1976. Biogeography of free-living soil nematodes from the perspective of plate tectonics. *Science,* 193: 508 – 510.

Findley, J.S., 1967. Insectivores and dermopterans. In: *Recent Mammals of the World* (S. Anderson and J.K. Jones, eds.). Roland Press, New York, N.Y., pp. 87 – 108.

Fischer, A.G., 1981. Climatic oscillations in the biosphere. In: *Biotic Crises in Ecological and Evolutionary Time* (M.H. Nitecki, ed.). Academic Press, New York, N.Y., pp. 103 – 131.

Fischer, A.G., 1984. The two Phanerozoic supercycles. In: *Catastrophes and Earth History* (W.A. Berggren and J.A. Van Couvering, eds.). Princeton Universtiy Press, Princeton, N.J., pp. 129 – 150.

Fischer-Piette, E. and Blanc, F., 1972. Le peuplement de mollusques terrestres de Madagascar. In: *Biogeography and Ecology in Madagascar* (R. Battistini and G. Richard-Vindard, eds.). W. Junk, The Hague, pp. 459 – 476.

Fleming, C.A., 1975. The geological history of New Zealand and its biota. In: *Biogeography and Ecology in New Zealand* (G. Kuschel, ed.). W. Junk, The Hague, pp. 1 – 86.

Flessa, K.W., 1980. The biological effects of plate tectonics and continental drift. *BioScience,* 30: 518 – 523.

Flint, O.S., Jr., 1978. Probable origins of the West Indian Trichoptera and Odonata faunas. In: *Proceedings of the 2nd International Symposium on Trichoptera* (M.I. Crichton, ed.). W. Junk, The Hague, pp. 215 – 223.

Flores, G., 1970. Suggested origin of the Mozambique Channel. *Trans. Geol. Soc. S. Afr.*, 73: 1 – 16.

Florin, R., 1963. The distribution of conifer and taxad genera in time and space. *Acta Horti Bergiani,* 20: 121 – 312.

Forbes, E., 1859. *The Natural History of European Seas* – edited and continued by Robert Goodwin-Austen. John Van Voorst, London.

Forster, R.R., 1975. The spiders and harvestmen. In: *Biogeography and Ecology in New Zealand* (G. Kuschel, ed.). W. Junk, The Hague, pp. 493–505.

Fujita, K., 1978. Pre-Cenozoic tectonic evolution of northeast Siberia. *J. Geol.*, 86: 159–172.

Gabunia, L.K. and Shevyreva, N.S., 1985. First marsupial fossil from Asia. *Dokl. Acad. Nauk SSSR*, 281: 685 (in Russian).

Galloway, D.J., 1979. Biogeographical elements in the New Zealand lichen flora. In: *Plants and Islands* (D. Bramwell, ed.). Academic Press, London, pp. 201–224.

Gansser, A., 1964. *Geology of the Himalayas.* Wiley-Interscience, New York, N.Y.

George, W., 1981. Wallace and his line. In: *Wallace's Line and Plate Tectonics* (T.C. Whitmore, ed.). Clarendon Press, Oxford, pp. 3–8.

Gilbert, C.R., 1976. Composition and derivation of the North American freshwater fish fauna. *Fla. Sci.*, 39: 104–111.

Gingerich, P.D., 1985. South American mammals in the Paleocene of North America. In: *The Great American Interchange* (F.G. Stehli and S.D. Webb, eds.). Plenum Press, New York, N.Y., pp. 123–137.

Glaessner, M.F., 1960. The fossil decapod Crustacea of New Zealand and the evolution of the order Decapoda. *N.Z. Geol. Surv. Palaeontol. Bull.*, 31: 1–63.

Godley, E.J., 1975. Flora and vegetation. In: *Biogeography and Ecology in New Zealand* (G. Kuschel, ed.). W. Junk, The Hague, pp. 177–229.

Goin, C.J. and Goin, O.B., 1971. *Introduction to Herpetology.* W.H. Freeman and Co., San Fransisco, Calif.

Goldblatt, P., 1978. An analysis of the flora of southern Africa: its characteristics, relationships, and origins. *Ann. Mo. Bot. Gard.*, 65: 369–436.

Good, R., 1974. *The Geography of the Flowering Plants.* Longman, London.

Gould, S.J., 1980. *The Panda's Thumb.* W.W. Norton, New York, N.Y.

Grande, L., 1985. The use of paleontology in systematics and biogeography, and a time control refinement for historical biogeography. *Paleobiology,* 11: 234–243.

Gray, A., 1859. Diagnostic characters of new species of phanerogamous plants collected in Japan by Charles Wright, botanist of the U.S. North Pacific Exploring Expedition. *Mem. Am. Acad. Arts,* 6: 377–452.

Greer, A.E., 1970. A subfamilial classification of scincid lizards. *Bull. Mus. Comp. Zool., Harvard Univ.,* 139: 151–183.

Grekoff, N. and Krommelbein, K., 1967. Etude comparée des ostracodes Mésozoiques continentaux des bassins Atlantiques: Série de Cocobeach, Gabon et Série de Bahia, Bresil. *Rev. Inst. Fr. Pet., Paris,* 22: 1307–1353.

Hallam, A., 1981. Relative importance of plate movements, eustasy, and climate in controlling major biogeographical changes since the early Mesozoic. In: *Vicariance Biogeography: A Critique* (G. Nelson and D.E. Rosen, eds.). Columbia, University Press, New York, N.Y., pp. 303–330.

Hand, S., 1984. Australia's oldest rodents: master mariners from Malaysia. In: *Vertebrate Zoogeography and Evolution in Australasia* (M. Archer and G. Clayton, eds.). Hesperian Press, Carlisle, W. Aust., pp. 905–919.

Harrison, A.D., 1978. Freshwater invertebrates (except molluscs). In: *Biogeography and Ecology of Southern Africa* (M.J.A. Werger, ed.). W. Junk, The Hague, pp. 1139–1152.

Hayes, D.E. and Ringis, J., 1973. Sea floor spreading in the Tasman Sea. *Nature,* 243: 454–458.

Hecht, M.K. and Archer, M., 1977. Presence of xiphodont crocodilians in the Tertiary and Pleistocene of Australia. *Aleheringa,* 10: 383–385.

Hedges, S.B., 1983. Caribbean biogeography: implications of recent plate tectonic studies. *Syst. Zool.,* 31: 518–522.

Heim de Balsac, H., 1972. Insectivores. In: *Biogeography and Ecology in Madagascar* (R. Battistini and G. Richard-Vindard, eds.). W. Junk, The Hague, pp. 629–660.

Herman, Y. and Hopkins, D.M., 1980. Arctic Ocean climate in late Cenozoic time. *Science,* 209: 557–562.

Hershkowitz, P., 1972. The recent mammals of the Neotropical region: a zoogeographical and ecological review. In: *Evolution, Mammals and Southern Continents* (A. Keast, F.C. Erk and B. Glass, eds.). State University of New York Press, Albany, N.Y., pp. 331–431.

Heyer, W.R., 1975. A preliminary analysis of the intergeneric relationships of the frog family Leptodactylidae. *Smithson. Contrib. Zool.,* 199: 1 – 55.

Hickey, L.J., 1984. Changes in the angiosperm flora across the Cretaceous-Tertiary boundary. In: *Catastrophes and Earth History* (W.A. Berggren and J.A. Van Couvering, eds.). Princeton University Press, Princeton, N.J., pp. 279 – 313.

Hoch, E., 1983. Fossil evidence of early Tertiary North Atlantic events viewed in European context. In: *Structure and Development of the Greenland-Scotland Ridge* (M.H.P. Bott, S. Saxon, M. Talwani and J. Thiede, eds.). Plenum Press, New York, N.Y., pp. 401 – 415.

Hoffstetter, R., 1980. Origin and deployment of the New World monkeys emphasizing the southern continents route. In: *Evolutionary Biology of the New World Monkeys and Continental Drift* (R.L. Ciochon and A.P. Chiarelli, eds.). Plenum Press, New York, N.Y., pp. 103 – 122.

Holloway, J.D., 1974. The biogeography of Indian butterflies. In: *Ecology and Biogeography in India* (M.S. Mani, ed.). W. Junk, The Hague, pp. 473 – 499.

Holloway, J.D., 1979. *A Survey of the Lepidoptera, Biogeography and Ecology of New Caledonia.* W. Junk, The Hague.

Hooker, J.D., 1853. *The Botany of the Antarctic Voyage of H.M. Discovery Ships Erebus and Terror in the Years 1839 – 1843, II. Flora Novae-Zelandiae.* Lovell Reeve, London.

Hopkins, D.M., 1967. The Cenozoic history of Beringia – a synthesis. In: *The Bering Land Bridge* (D.M. Hopkins, ed.). Stanford University Press, Stanford, Calif., pp. 451 – 484.

Hora, S.L., 1937. Geographical distribution of Indian freshwater fishes and its bearing on the probable land connections between India and adjacent countries. *Curr. Sci. Bangalore,* 5: 351 – 356.

Hotchkiss, F.H.C., 1982. Antarctic fossil echinoids: review and current research. In: *Antarctic Geoscience* (C. Craddock, ed.). University of Wisconsin Press, Madison, Wisc., pp. 679 – 684.

Howden, H.F., 1981. Zoogeography of some Australian Coleoptera as exemplified by the Scarabaeoides. In: *Ecological Biogeography of Australia* (A. Keast, ed.). W. Junk, The Hague, pp. 1009 – 1035.

Howden, H.F. and Cooper, J.B., 1977. The generic classification of the Bolboceratini of the Australian region, with descriptions of four new genera (Scarabeidae: Geotrupinae). *Aust. J. Zool., Suppl. Ser.,* 50: 1 – 50.

Hubbell, T.H., 1968. The biology of islands. *Proc. Natl. Acad. Sci. U.S.A.,* 60: 22 – 32.

Humphries, C.J., 1981. Biogeographical methods and the southern beeches. In: *The Evolving Biosphere* (P.L. Forey, ed.). Cambridge University Press, pp. 283 – 297.

Humphries, C.J. and Parenti, L.R., 1986. *Cladistic Biogeography.* Clarendon Press, Oxford.

Hull, F.M., 1962. Robber flies of the world, 1. *Bull. U.S. Nat. Mus.,* 224: 1 – 430.

Illies, J., 1968. Biogeography and ecology of Neotropical freshwater insects, especially those from running waters. In: *Biogeography and Ecology in South America* (E.J. Fittkau et al., eds.). W. Junk, The Hague, pp. 685 – 708.

Imbrie, J. and Imbrie, K.P., 1979. *Ice Ages.* Enslow, N.J.

Jaeckel, S.G.A., 1968. Die mollusken Sudamerikas. In: *Biogeography and Ecology in South America* (E.J. Fittkau et al., eds.). W. Junk, The Hague, pp. 794 – 827.

Jamieson, B.G.M., 1981. Historical biogeography of Australian Oligochaeta. In: *Ecological Biogeography of Australia* (A. Keast, ed.). W. Junk, The Hague, pp. 887 – 921.

Jardine, N. and McKenzie, O., 1972. Continental drift and the dispersal and evolution of organisms. *Nature,* 235: 20 – 24.

Jayaram, K.C., 1974. Ecology and distribution of freshwater fishes, Amphibia, and reptiles. In: *Ecology and Biogeography in India* (M.S. Mani, ed.). W. Junk, The Hague, pp. 517 – 584.

Jeekel, C.A.W., 1974. The group taxonomy and geography of the Sphaerotheriidae (Diplopoda). In: *Myriapoda* (J.G. Blower, ed.). *Symp. Zool. Soc. London,* 32: 41 – 52.

Johnson, B.D., Powell, C.M. and Veevers, J.J., 1976. Spreading history of the eastern Indian Ocean and greater India's northward flight from Antarctica and Australia. *Geol. Soc. Am. Bull.,* 87: 1560 – 1566.

Johnson, L.A.S. and Briggs, B.G., 1975. On the Proteaceae: the evolution and classification of a southern family. *Bot. J. Linn. Soc.,* 70: 83 – 182.

Kalandadze, N.N., 1975. The first discovery of *Lystrosaurus* in the European regions of the U.S.S.R. *Paleontol. Zhur.,* 4: 140 – 142 (in Russian).

Kauffman, E.G., 1973. Cretaceous bivalvia. In: *Atlas of Paleobiogeography* (A. Hallam, ed.). Elsevier, Amsterdam, pp. 353 – 383.

Kauffman, E.G., 1979. Cretaceous. In: *Treatise on Invertebrate Paleontology, Part A. Introduction.* (R.A. Robison and C. Teichert, eds.). Geological Society of America/University of Kansas, pp. 418 – 487.

Kay, E.A., 1979. *Little Worlds of the Pacific: An Essay on Pacific Basin Biogeography.* Harold L. Lyon Aboretum, University of Hawaii, Honolulu, Hawaii.

Kay, E.A., 1984. Patterns of speciation in the Indo-West Pacific. In: *Biogeography of the Tropical Pacific* (F.J. Radovsky, P.H. Raven and S.H. Sohmer, eds.). Assoc. Syst. Coll. and Bishop Museum, pp. 15 – 31.

Keast, J.A., 1981. The evolutionary biogeography of Australian birds. In: *Ecological Biogeography of Australia* (A. Keast, ed.). W. Junk, The Hague, pp. 1587 – 1635.

Keast, J.A., 1983. In the steps of Alfred Russel Wallace: biogeography of the Asian-Australian interchange zone. In: *Evolution, Time and Space: The Emergence of the Biosphere* (R.W. Sims, J.H. Price and P.E.S. Whalley, eds.). Academic Press, London, pp. 367 – 407.

Kennett, J.P., 1982. *Marine Geology.* Prentice-Hall, Englewood Cliffs, N.J.

Kiener, A. and Richard-Vindard, G., 1972. Fishes of the continental waters of Madagascar. In: *Biogeography and Ecology in Madagascar* (R. Battistini and G. Richard-Vindard, eds.). W. Junk, The Hague, pp. 477 – 499.

King, L., 1980. Gondwanaland reunited. *Geology,* 8: 111 – 112.

Kirsch, J.A.W., 1977. The comparative serology of Marsupialia, and a classification of marsupials. *Aust. J. Zool., Suppl. Ser.,* 52: 1 – 152.

Kluge, A.G., 1967. Higher taxonomic categories of gekkonid lizards and their evolution. *Bull. Am. Mus. Nat. Hist.,* 135: 1 – 60.

Knudsen, J., 1970. The systematics and biology of abyssal and hadal Bivalvia. *Galathea Rep.,* 11: 7 – 236.

Koch, L.E., 1981. The scorpions of Australia: aspects of their ecology and zoogeography. In: *Ecological Biogeography of Australia* (A. Keast, ed.). W. Junk, The Hague, pp. 875 – 884.

Koechlin, J., 1972. Flora and vegetation of Madagascar. In: *Biogeography and Ecology in Madagascar* (R. Battistini and G. Richard-Vindard, eds.). W. Junk, The Hague, pp. 145 – 190.

Krassilov, V.A., 1972. Phytogeographical classification of Mesozoic floras and their bearing on continental drift. *Nature,* 237: 49 – 50.

Krishna, K., 1970. Taxonomy, phylogeny, and distribution of termites. In: *Biology of Termites, Vol. II* (K. Krishna and F.M. Weesner, eds.). Academic Press, New York, N.Y., pp. 127 – 152.

Krishnan, M.S., 1974. Geology. In: *Ecology and Biogeography in India* (M.S. Mani, ed.). W. Junk, The Hague, pp. 60 – 98.

Krommelbein, K., 1979. African Cretaceous ostracods and their relations to surrounding continents. In: *Historical Biogeography, Plate Tectonics and the Changing Environment* (J. Gray and A.J. Boucot, eds.). Oregon State University Press, Corvallis, Oreg., pp. 305 – 310.

Kurtén, B., 1973. Early Tertiary land mammals. In: *Atlas of Paleobiogeography* (A. Hallam, ed.). Elsevier, Amsterdam, pp. 437 – 442.

Kurtén, B. and Anderson, E., 1980. *Pleistocene Mammals of North America.* Columbia University Press, New York, N.Y.

Kurup, G.V., 1974. Mammals of Assam and the mammal-geography of India. In: *Ecology and Biogeography in India* (M.S. Mani, ed.). W. Junk, The Hague, pp. 585 – 613.

Lack, D., 1976. *Island Biology as Illustrated by the Land Birds of Jamaica. Studies in Ecology, Vol. 3.* University of California Press, Berkeley, Calif.

Lansbury, I., 1981. Aquatic and semi-aquatic bugs (Hemiptera) of Australia. In: *Ecological Biogeography of Australia* (A. Keast, ed.). W. Junk, The Hague, pp. 1197 – 1211.

Larson, R.L., 1977. Early Cretaceous breakup of Gondwanaland off western Australia. *Geology,* 5: 57 – 60.

Lasaga, A.C., Berner, R.A. and Garrels, R.M., 1985. An improved geochemical model of atmospheric CO_2 fluctuations over the past 100 million years. In: *The Carbon Cycle and Atmospheric CO_2:*

Natural Variations from Archean to Present (E.T. Sundaquist and W.S. Broecker, eds.). *Am. Geophys. Union, Geophys. Monogr.,* 32, pp. 397 – 411.

Laurent, R.F., 1979. Herpetofaunal relationships between Africa and South America. In: *The South American Herpetofauna: Its Origin, Evolution, and Dispersal* (W.E. Duellman, ed.). *Mus. Nat. Hist., Univ. Kansas, Monogr.,* 7: 55 – 71.

Lavocat, R., 1980. The implications of rodent paleontology and biogeography to the geographical sources and origin of platyrrhine primates. In: *Evolutionary Biology of the New World Monkeys and Continental Drift* (R.L. Ciochon and A.B. Chiarelli, eds.). Plenum Press, New York, N.Y., pp. 93 – 102.

Legendre, R., 1972. Les arachnides de Madagascar. In: *Biogeography and Ecology in Madagascar* (R. Battistini and G. Richard-Vindard, eds.). W. Junk, The Hague, pp. 427 – 457.

Leroy, J.R., 1978. Composition, origin and affinities of the Madagascan vascular flora. *Ann. Mo. Bot. Gard.,* 65: 535 – 589.

Li, H., 1952. Floristic relationships between eastern Asia and eastern North America. *Trans. Am. Philos. Soc., New Ser.,* 42: 371 – 429.

Lillegraven, J.A., Kraus, M.J. and Brown, T.M., 1979. Paleogeography of the world of the Mesozoic. In: *Mesozoic Mammals* (J.A. Lillegraven, Z. Kielan-Jaworowska and W.A. Clemens, eds.). University of Calif. Press, Berkeley, Calif., pp. 277 – 308.

Little, E.L., 1983. North American trees with relationships in eastern Asia. *Ann. Mo. Bot. Gard.,* 70: 605 – 615.

Lorence, D.H., 1985. A monograph of the Monimiaceae (Laurales) in the Malagasy region (southwest Indian Ocean). *Ann. Mo. Bot. Gard.,* 72: 1 – 165.

Lowe-McConnell, R.H., 1975. *Fish Communities in Tropical Freshwater.* Longman, London.

MacFadden, B.J., 1980. Rafting mammals or drifting islands? Biogeography of the Greater Antillean insectivores *Nesophontes* and *Solenodon. J. Biogeogr.,* 7: 11 – 22.

MacFadden, B.J., 1985. Patterns of phylogeny and rates of evolution in fossil horses: hipparions from the Miocene and Pliocene of North America. *Paleobiology,* 11: 245 – 257.

Mackerras, I.M., 1970. Composition and distribution of the fauna. In: *The Insects of Australia.* Melbourne University Press, Parkville, Vic., pp. 187 – 203.

Maglio, V.J., 1978. Patterns of faunal evolution. In: *Evolution of African Mammals* (V.J. Maglio and H.B.S. Cooke, eds.). Harvard University Press, Cambridge, Mass., pp. 603 – 619.

Mahe, J., 1972. The Malagasy subfossils. In: *Biogeography and Ecology in Madagascar* (R. Battistini and G. Richard-Vindard, eds.). W. Junk, The Hague, pp. 339 – 365.

Main, B.Y., 1981. Australian spiders: diversity, distribution and ecology. In: *Ecological Biogeography of Australia* (A. Keast, ed.). W. Junk, The Hague, pp. 809 – 852.

Malfait, B.T. and Dinkelman, M.G., 1972. Circum-Caribbean tectonic and igneous activity and the evolution of the Caribbean plate. *Geol. Soc. Am. Bull.,* 83: 251 – 272.

Mani, M.S., 1974a. Biogeography of the peninsula. In: *Ecology and Biogeography in India* (M.S. Mani, ed.). W. Junk, The Hague, pp. 614 – 647.

Mani, M.S., 1974b. The flora. In: *Ecology and Biogeography in India* (M.S. Mani, ed.). W. Junk, The Hague, pp. 159 – 177.

Marincovich, L., Brouwers, E.M. and Carter, L.D., 1985. Early Tertiary marine fossils from northern Alaska: implications for Arctic Ocean paleogeography and faunal evolution. *Geology,* 13: 770 – 773.

Marshall, L.G., 1981. The great American interchange – an invasion induced crisis for South American mammals. In: *Biotic Crises in Ecological and Evolutionary Time* (M.H. Nitecki, ed.). Academic Press, New York, N.Y., pp. 133 – 229.

Marshall, L.G., Webb, S.D., Sepkoski, J.J., Jr. and Raup, D.M., 1982. Mammalian evolution and the great American interchange. *Science,* 215: 1351 – 1357.

Marx, H. and Rabb, G.B., 1965. Relationships and zoogeography of the viperine snakes. *Fieldiana, Zool.,* 44: 161 – 206.

Matsumoto, T., 1973. Late Cretaceous Ammonoidea. In: *Atlas of Paleobiogeography* (A. Hallam, ed.). Elsevier, Amsterdam, pp. 421 – 429.

Matthew, W.D., 1915. Climate and evolution. *Ann. N.Y. Acad. Sci.,* 24: 171 – 318.

Matthew, W.D., 1918. Affinities and origin of the Antillean mammals. *Geol. Soc. Am. Bull.,* 29: 657 – 666.

Mattson, P.H., 1984. Caribbean structural breaks and plate movements. In: *The Caribbean – South American Plate Boundary and Regional Tectonics* (W.E. Bonini, R.B. Hargraves and R. Shagam, eds.). *Geol. Soc. Am., Mem.,* 162: 131 – 152.

Maxson, L.R., 1984. Molecular probes of phylogeny and biogeography in toads of the widespread genus *Bufo. Mol. Biol. Evol.,* 1: 345 – 356.

Maxson, L.R. and Wilson, A.C., 1975. Albumin evolution and organismal evolution in tree frogs (Hylidae). *Syst. Zool.,* 24: 1 – 15.

Mayr, E., 1946. History of North American bird fauna. *Wilson Bull.,* 58: 3 – 41.

Mayr, E., 1976. History of the North American bird fauna. In: *Evolution and the Diversity of Life* (E. Mayr, ed.). Harvard University Press, Cambridge, Mass., pp. 565 – 588.

McCoy, E.D. and Heck, K.L., 1983. Centers of origin revisited. *Paleobiology,* 9: 17 – 19.

McDowall, R.M., 1978. *New Zealand Freshwater Fishes.* Heinemann, Auckland.

McKenna, M.C., 1975. Fossil mammals and the Early Eocene North Atlantic land community. *Ann. Mo. Bot. Gard.,* 62: 335 – 353.

McKenna, M.C., 1980. Early history and biogeography of South America's extinct land mammals. In: *Evolutionary Biology of the New World Monkeys and Continental Drift* (R.L. Ciochon and A.B. Chiarelli, eds.). Plenum Press, New York, N.Y., pp. 43 – 77.

McKenna, M.C., 1983a. Cenozoic paleogeography of North Atlantic land bridges. In: *Structure and Development of the Greenland – Scotland Ridge* (M.H.P. Bott, S. Saxov, M. Talwani and J. Thiede, eds.). Plenum Press, New York, N.Y., pp. 351 – 399.

McKenna, M.C., 1983b. Holarctic landmass rearrangement, cosmic events, and Cenozoic terrestrial organisms. *Ann. Mo. Bot. Gard.,* 70: 459 – 489.

McLellan, I.D., 1975. The freshwater insects. In: *Biogeography and Ecology in New Zealand* (G. Kuschel, ed.). W. Junk, The Hague, pp. 537 – 559.

McLellan, I.D., 1977. New alpine and southern Plecoptera from New Zealand, and a new classification of the Gripopterygidae. *N.Z. J. Zool.,* 4: 119 – 147.

McMichael, D.F., 1958. The nature and origin of the New Zealand freshwater mussel fauna. *Trans. R. Soc. N.Z.,* 85: 427 – 432.

McMichael, D.F., 1967. Australian freshwater Mollusca and their probable evolutionary relationships: a summary of present knowledge. In: *Australian Inland Waters and Their Fauna* (A.H. Weatherle, ed.). Australian National University Press, Canberra, A.C.T., pp. 123 – 149.

Menon, A.G.K., 1955. The external relationships of the Indian freshwater fishes, with special reference to the countries bordering on the Indian Ocean. *J. Asiat. Soc. Sci.,* 21: 31 – 38.

Menzies, R.J., 1965. Conditions for the existence of life on the abyssal sea floor. *Mar. Biol. Ann. Rev.,* 3: 195 – 210.

Michener, C.D., 1979. Biogeography of the bees. *Ann. Mo. Bot. Gard.,* 66: 277 – 347.

Miller, R.R., 1982. Pisces. In: *Aquatic Biota of Mexico, Central America and the West Indies* (S.H. Hurlbert and A. Villalobos-Figueroa, eds.). San Diego State University Press, San Diego, Calif., pp. 486 – 501.

Milner, A.R., 1983. The biogeography of salamanders in the Mesozoic and early Cenozoic: a cladistic-vicariance model. In: *Evolution, Time and Space: The Emergence of the Biosphere* (R.W. Sims, J.H. Price and P.E.S. Whalley, eds.). Academic Press, New York, N.Y., pp. 431 – 468.

Molnar, R.E., 1981. A dinosaur from New Zealand. In: *Gondwana Five* (M.M. Cresswell and P. Vella, eds.). *Proc. 5th Int. Gondwana Symp., Wellington,* pp. 91 – 96.

Molnar, R.E., 1984. Palaeozoic and Mesozoic reptiles and amphibians from Australia. In: *Vertebrate Zoogeography and Evolution in Australasia* (M. Archer and G. Clayton, eds.). Hesperian Press, Carlisle, W. Aust., pp. 331 – 336.

Monroe, W.H., 1980. Geology of the middle Tertiary formations of Puerto Rico. *U.S. Geol. Surv., Prof. Pap.,* 953: 1 – 93.

Moore, D.M., 1972. Connections between cool temperate floras with particular reference to southern South America. In: *Taxonomy, Phytogeography and Evolution* (D.H. Valentine, ed.). Academic Press, New York, N.Y., pp. 115 – 138.

Moore, H.E., Jr., 1973. Palms in the tropical forest ecosystems of Africa and South America. In: *Tropical Forest Ecosystems in Africa and South America: A Comparative Review* (B.J. Dreggers, E.S. Ayensu and W.D. Duckworth, eds.). Smithsonian Press, Washington, D.C., pp. 63 – 88.

Mound, L.A., 1972. Further studies on Australian Aeolothripidae (Thysanoptera). *J. Aust. Entomol. Soc.,* 11: 37 – 54.

Muller, J., 1970. Palynological evidence on early differentiation of angiosperms. *Biol. Rev.,* 45: 417 – 450.

Murray, J., 1895. A summary of the scientific results. In: *Challenger Report Summary.* 2 volumes, London.

Myers, G.S., 1938. Freshwater fishes and West Indian zoogeography. *Smithson. Rep.,* 1937: 339 – 364.

Myers, G.S., 1958. Trends in the evolution of teleostean fishes. *Stanford Ichthy. Bull.,* 7: 27 – 30.

Myers, G.S., 1966. Derivation of the freshwater fish fauna of Central America. *Copeia,* (4): 766 – 773.

Neboiss, A., 1977. A taxonomic and zoogeographic study of Tasmanian caddis-flies (Insecta: Trichoptera). *Mem. Nat. Mus. Vic.,* 38: 1 – 208.

Nelson, G., 1978. From Candolle to Croizat: comments on the history of biogeography. *J. Hist. Biol.,* 11: 269 – 305.

Nelson, G. and Platnick, N.I., 1980. A vicariance approach to historical biogeography. *BioScience,* 30: 339 – 343.

Nelson, G. and Platnick, N., 1984. Biogeography. *Carolina Biol. Readers* 119: 1 – 16.

Nelson, G. and Rosen, D.E. (eds.), 1981. *Vicariance Biogeography: A Critique.* Columbia University Press, New York, N.Y.

Nelson, J.S., 1984. *Fishes of the World.* John Wiley and Sons, New York, N.Y.

Niklas, K.J., 1981. Discussion. In: *Vicariance Biogeography: A Critique* (G. Nelson and D.E. Rosen, eds.). Columbia University Press, New York, N.Y., pp. 428 – 435.

Nunns, A.G., 1983. Plate tectonic evolution of the Greenland – Scotland Ridge and surrounding regions. In: *Structure and Development of the Greenland – Scotland Ridge* (M.H.P. Bott, S. Saxov, M. Talwani and J. Thiede, eds.). Plenum Press, New York, N.Y., pp. 11 – 30.

Nussbaum, R.A., 1984. Amphibians of the Seychelles. In: *Biogeography and Ecology of the Seychelles Islands* (D.R. Stoddart, ed.). W. Junk, The Hague, pp. 379 – 415.

Obruchev, D.V., 1964. *Fundamentals of Paleontology, Vol. XI. Agnatha, Pisces.* USSR Academy of Sciences, Moscow (English translation by Israel Program for Scientific Translations, Jerusalem, 1967).

Olson, E.C., 1979. Biological and physical factors in the dispersal of Permo-Carboniferous terrestrial vertebrates. In: *Historical Biogeography, Plate Tectonics and the Changing Environment* (J. Gray and A.J. Boucot, eds.). Oregon State University Press, Corvallis, Oreg., pp. 227 – 238.

Owen, H.G., 1983. *Atlas of Continental Displacement, 200 Million Years to the Present.* Cambridge University Press, Cambridge.

Padian, K. and Clemens, W.A., 1985. Terrestrial vertebrate diversity: episodes and insights. In: *Phanerozoic Diversity Patterns* (J.W. Valentine, ed.). Princeton University Press, Princeton, N.J., pp. 41 – 96.

Paramonov, S.J., 1959. Zoogeographical aspects of the Australian dipterofauna. In: *Biogeography and Ecology in Australia* (A. Keast, R.L. Crocker and C.S. Christian, eds.). W. Junk, The Hague.

Parenti, L.R., 1981. A phylogenetic and biogeographic analysis of Cyprinodontiform fishes (Teleostei, Atherinomorpha). *Bull. Am. Mus. Nat. Hist.,* 168: 335 – 557.

Parker, H.W., 1977. *Snakes: A Natural History* (2 nd ed., rev. by A.G.C. Grandison). Cornell University Press, Ithaca.

Parodiz, J.J. and Bonetto, A.A., 1962. Taxonomy and zoogeographic relationships of the South American Naiades (Pelecypoda: Unionacea and Mutelacea). *Malacologia,* 1: 179 – 213.

Patriat, P. and Achache, J., 1984. India-Eurasia collision chronology has implications for crustal shortening and driving mechanism for plates. *Nature,* 311: 615 – 621.

Patterson, C., 1975. The distribution of Mesozoic freshwater fishes. *Mem. Mus. Nat. Hist. Nat.,* Paris, 88: 156 – 173.

Patterson, C., 1981a. The development of the North American fish fauna – a problem of historical biogeography. In: *The Evolving Biosphere* (P.L. Lorey, ed.). Cambridge University Press, Cambridge, pp. 265 – 281.

Patterson, C., 1981b. Methods of paleobiogeography. In: *Vicariance Biogeography: A Critique* (G. Nelson and D.E. Rosen, eds.). Columbia University Press, New York, N.Y., pp. 446 – 489.

Paulain, R., 1970. The termites of Madagascar. In: *Biology of Termites, Vol. II* (K. Krishna and F.M. Weesner, eds.). Academic Press, New York, N.Y., pp. 281 – 294.

188

Paulain, R., 1972. Some ecological and biogeographical problems of the entomofauna of Madagascar. In: *Biogeography and Ecology in Madagascar* (R. Battistini and G. Richard-Vindard, eds.). W. Junk, The Hague, pp. 411 – 426.

Peake, J., 1978. Distribution and ecology of the Stylommatophora. In: *Pulmonates, Vol. 2A. Systematics, Evolution and Ecology* (V. Fretter and J. Peake, eds.). Academic Press, New York, N.Y., pp. 429 – 526.

Pescador, M.L. and Peters, W.L., 1980. Phylogenetic relationships and zoogeography of cool-adapted Leptophlebiidae (Ephemeroptera) in southern South America. In: *Advances in Ephemeroptera Biology* (J.F. Flannagan and K.E. Marshall, eds.). Plenum Press, New York, N.Y., pp. 43 – 56.

Petter, F., 1972. The rodents of Madagascar: the seven genera of Malagasy rodents. In: *Biogeography and Ecology in Madagascar* (R. Battistini and G. Richard-Vindard, eds.). W. Junk, The Hague, pp. 661 – 665.

Petter, J.J., 1972. Order of primates: sub-order of lemurs. In: *Biogeography and Ecology in Madagascar* (R. Battistini and G. Richard-Vindard, eds.). W. Junk, The Hague, pp. 683 – 702.

Phipps, J.B., 1983. Biogeographic, taxonomic, and cladistic relationships between East Asiatic and North American *Crataegus. Ann. Mo. Bot. Gard.,* 70: 667 – 700.

Pindell, J. and Dewey, J.F., 1982. Permo-Triassic reconstruction of western Pangea and the evolution of the Gulf of Mexico/Caribbean region. *Tectonics,* 1: 179 – 211.

Pinkey, E., 1978. Odonata. In: *Biogeography and Ecology of Southern Africa* (M.J.A. Werger, ed.). W. Junk, The Hague, pp. 723 – 731.

Platnick, N.I., 1976. Drifting spiders or continents? Vicariance biogeography of the spider family Laroniinae (Araneae: Gnaphosidae). *Syst. Zool.,* 25: 101 – 109.

Pregill, G.K., 1981. An appraisal of the vicariance hypothesis of Caribbean biogeography and its application to West Indian terrestrial vertebrates. *Syst. Zool.,* 30: 147 – 155.

Procter, J., 1984. Floristics of the granitic islands of the Seychelles. In: *Biogeography and Ecology of the Seychelles Islands* (D.R. Stoddart, ed.). W. Junk, The Hague, pp. 209 – 220.

Rabb, G.B and Marx, H., 1973. Major ecological and geographic patterns in the evolution of colubroid snakes. *Evolution,* 27(1): 69 – 83, 4 figs.

Rabinowitz, P.D., Coffin, M.L. and Falvey, D., 1983. The separation of Madagascar and Africa. *Science,* 220: 67 – 69.

Randall, J.E., 1982. Examples of antitropical and antiequatorial distribution of Indo-West Pacific fishes. *Pac. Sci.,* 35: 197 – 209.

Raup, D.M., 1979. Size of the Permo-Triassic bottleneck and its evolutionary implications. *Science,* 206: 217 – 218.

Raup, D.M. and Sepkoski, J.J., Jr., 1984. Periodicity of extinctions in the geologic past. *Proc. Natl. Acad. Sci.,* 81: 801 – 805.

Raven, P.H. and Axelrod, D.I., 1972. Plate tectonics and Australasian paleobiogeography. *Science,* 176: 1379.

Raven, P.H. and Axelrod, D.I., 1974. Angiosperm biogeography and past continental movements. *Ann. Mo. Bot. Gard.,* 61: 539 – 673.

Raven, P.H. and Axelrod, D.I., 1975. History of the flora and fauna of Latin America. *Am. Sci.,* 63: 420 – 429.

Rawson, P.F., 1981. Early Cretaceous ammonite biostratigraphy and biogeography. In: *The Ammonoidea* (M.R. House and J.R. Senior, eds.). Academic Press, London, pp. 499 – 529.

Reichardt, H., 1979. The South American carabid fauna: endemic tribes and tribes with African relationships. In: *Carabid Beetles: Their Evolution, Natural History, and Classification* (T.L. Erwin et al., eds.). W. Junk, The Hague, pp. 319 – 325.

Rentz, D.C., 1978. Orthoptera. In: *Biogeography and Ecology of Southern Africa* (M.J.A. Werger, ed.). W. Junk, The Hague, pp. 733 – 746.

Repenning, C.A., 1980. Faunal exchange between Siberia and North America. In: *Proceedings, 5th Biennial Conference of the American Quaternary Association. Can. J. Anthropol.,* 1: 37 – 44.

Rich, P.V., 1975. Antarctic dispersal routes, wandering continents, and the origin of Australian non-passeriform avifauna. *Mem. Nat. Mus. Vic.,* 36: 63 – 126.

Rich, P.V., 1979. Fossil birds of old Gondwanaland: a comment on drifting continents and their passengers. In: *Historical Biogeography, Plate Tectonics, and the Changing Environment* (J. Gray and A.J. Boucot, eds.). Oregon State University Press, Corvallis, Oreg., pp. 321 – 332.

Rich, P.V., 1980. The Australian Dromorthinidae: a group of extinct large ratites. *Contrib. Sci. Nat. Hist. Mus. Los Angeles County,* 330: 93 – 103.

Richards, P.W., 1973. Africa, the "odd man out". In: *Tropical Forest Ecosystems in Africa and South America: A Comparative Review* (B.J. Meggers, E.S. Ayensu and W.D. Duckworth, eds.). Smithsonian Press, Washington, D.C., pp. 21 – 26.

Riek, E.F., 1972. The phylogeny of the Parastacidae (Crustacea: Astacoidea), and description of a new genus of Australian freshwater crayfishes. *Aust. J. Zool.,* 20: 369 – 389.

Rigassi, D., 1963. Sur la Géologie de la Sierra de los Organos, Cuba. *Arch. Sci. (Genève), Soc. Phys. Hist. Nat.,* 16: 339 – 350.

Rivas, L.R., 1958. The origin, evolution, dispersal, and geographical distribution of the Cuban poeciliid fishes of the tribe Girardinini. *Proc. Am. Philos. Soc.,* 102: 281 – 320.

Robinson, E. and Lewis, J.F., 1971. Field guide to aspects of the geology of Jamaica. In: *International Field Institute Guidebook to the Caribbean Island-Arc System.* American Geological Institute, Falls Church, Va., pp. 2 – 29.

Rona, P.A., Bostrom, K., Laubier, L. and Smith, K.L., Jr. (eds.), 1983. *Hydrothermal Processes at Seafloor Spreading Centers. NATO Conference Series IV: Marine Sciences.* Plenum Press, New York, N.Y.

Rosa, D., 1923. Qu'est-ce que l'hologénèse? *Scientia,* 33: 113 – 124.

Rosen, D.E., 1976. A vicariance model of Caribbean biogeography. *Syst. Zool.,* 24: 431 – 464.

Rosen, D.E., 1985. Geological hierarchies and biogeographic congruence in the Caribbean. *Ann. Mo. Bot. Gard.,* 72: 636 – 659.

Rosen, D.E. and Bailey, R.M., 1963. The poeciliid fishes (Cyprinodontiformes), their structure, zoogeography, and systematics. *Bull. Am. Mus. Nat. Hist.,* 126: 1 – 176.

Ross, H.H., 1958. Affinities and origins of the northern and montane insects of western North America. In: *Zoogeography* (C.L. Hubbs, ed.). American Association for the Advancement of Science, Washington, D.C., pp. 231 – 252.

Ross, H.H., 1967. The evolution and past distribution of the Trichoptera. *Ann. Rev. Entomol.,* 12: 169 – 206.

Russell, D.A., 1984. Terminal Cretaceous exinctions of large reptiles. In: *Catastrophes and Earth History* (W.A. Berggren and J.A. Van Couvering, eds.). Princeton University Press, Princeton, N.J., pp. 373 – 384.

Sahni, A., 1984. Cretaceous-Paleocene terrestrial faunas of India: lack of endemism during drifting of the Indian plate. *Science,* 226: 441 – 443.

Sahni, A. and Kumar, V., 1974. Paleogene paleobiogeography of the Indian subcontinent. *Palaeogeogr., Palaeoclimatol., Palaeoecol.,* 15: 209 – 226.

Salvat, B., 1967. Importance de la fauna malacologique dans les atolls Polynésiens. *Cah. Pac.,* 11: 7 – 49.

Savage, D.E. and Russell, D.E., 1983. *Mammalian Paleofaunas of the World.* Addison-Wesley, Reading, Mass.

Savage, J.M., 1973. The geographic distribution of frogs: patterns and predictions. In: *Evolutionary Biology of the Anurans* (J.L. Vial, ed.). University of Missouri Press, Columbia, Mo., pp. 352 – 445.

Savage, J.M., 1982. The enigma of the Central American herpetofauna: dispersals or vicariance? *Ann. Mo. Bot. Gard.,* 69: 464 – 547.

Schaeffer, B., 1972. A Jurassic fish from Antarctica. *Am. Mus. Novit.,* 2495: 1 – 17.

Scharff, R.F., 1922. On the origin of the West Indian fauna. *Bijdr. Dierk.,* pp. 65 – 72.

Scheltema, R.S. and Williams, I.P., 1983. Long-distance dispersal of planktonic larvae and the biogeography and evolution of Polynesian and western Pacific molluscs. *Bull. Mar. Sci.,* 33: 545 – 565.

Schlinger, E.I., 1974. Continental drift, *Nothofagus,* and some ecologically associated insects. *Ann. Rev. Entomol.,* 19: 323 – 343.

Schminke, H.K., 1974. Mesozoic intercontinental relationships as evidenced by bathynellid Crustaceae (Syncarida: Malacostraca). *Syst. Zool.,* 23: 157 – 164.

Schopf, T.J.M., 1980. *Paleoceanography.* Harvard University Press, Cambridge, Mass.

Schuchert, C., 1935. *Historical Geology of the Antillean-Caribbean Region.* John Wiley, New York, N.Y.

190

Schuster, R.M., 1969. Problems of Antipodal distribution in lower land plants. *Taxon,* 18: 46 – 91.
Schuster, R.M., 1972. Continental movements, "Wallace's Line" and Indomalayan-Australasian dispersal of land plants: some eclectic concepts. *Bot. Rev.,* 38: 3 – 86.
Schuster, R.M., 1976. Plate tectonics and its bearing on the geographical origin and dispersal of angiosperms. In: *Origin and Early Evolution of Angiosperms* (C.B. Beck, ed.). Columbia University Press, New York, N.Y., pp. 48 – 138.
Schuster, R.M., 1981. Paleoecology, origin, distribution through time, and evolution of Hepaticae and Anthrocerotae. In: *Paleobotany, Paleoecology, and Evolution, Vol. 2* (K.J. Niklas, ed.). Praeger, New York, N.Y., pp. 129 – 191.
Sclater, J.G. and Tapscott, C., 1979. The history of the Atlantic. *Sci. Am.,* 240: 156 – 174.
Sclater, J.G., Hellinger, S. and Tapscott, C., 1977. The paleobathymetry of the Atlantic Ocean from the Jurassic to the present. *J. Geol.,* 85: 509 – 552.
Sclater, P.L., 1858. On the general geographical distribution of the members of the Class Aves. *J. Linn. Soc. (Zool.),* 2: 130 – 145.
Sen-Sarma, P.K., 1974. Ecology and biogeography of the termites of India. In: *Ecology and Biogeography in India* (M.S. Mani, ed.). W. Junk, The Hague, pp. 421 – 472.
Shields, O. and Dvorak, S.K., 1979. Butterfly distribution and continental drift between the Americas, the Caribbean and Africa. *J. Nat. Hist.,* 13: 221 – 250.
Sibley, C.G. and Ahlquist, J.E., 1986. Reconstructing bird phylogeny by comparing DNA's. *Sci. Am.,* 254: 82 – 92.
Sigé, B., 1972. La faunule de mammifères du Crétacé supérieur de Laguna Umayo (Andes péruviennes). *Mus. Natl. Hist. Nat. Bull., Ser. 3,* 19: 375 – 409.
Simons, E.L., 1960. The Paleocene Pantodonta. *Trans. Am. Philos. Soc.,* 50: 1 – 99.
Simpson, B.B. and Neff, J.L., 1985. Plants, their pollinating bees, and the great American interchange. In: *The Great American Interchange* (F.G. Stehli and S.D. Webb, eds.). Plenum Press, New York, N.Y., pp. 427 – 452.
Simpson, G.G., 1956. Zoogeography of West Indian land mammals. *Am. Mus. Novit.,* 1759: 1 – 28.
Simpson, G.G., 1965. *The Geography of Evolution.* Chilton Books, Philadelphia, Pa.
Simpson, G.G., 1980. *Splendid Isolation: The Curious History of South American Mammals.* Yale University Press, New Haven, Conn.
Sing-chi, C., 1983. A comparison of orchid floras of temperate North America and eastern Asia. *Ann. Mo. Bot. Gard.,* 70: 713 – 723.
Singh, S., 1974. Some aspects of the ecology and geography of Diptera. In: *Ecology and Biogeography in India* (M.S. Mani, ed.). W. Junk, The Hague, pp. 500 – 516.
Smiley, C.J., 1979. Pre-Tertiary phytogeography and continental drift-some apparent discrepancies. In: *Historical Biogeography, Plate Tectonics and the Changing Environment* (J. Gray and A.J. Boucot, eds.). Oregon State University Press, Corvallis, Oreg., pp. 311 – 319.
Smith, A.C., 1972. An appraisal of the orders and families of primitive extant angiosperms. *J. Indian Bot. Soc.,* 50A: 215 – 226.
Smith, A.C., 1973. Angiosperm evolution and the relationship of the floras of Africa and America. In: *Tropical Forest Ecosystems in Africa and South America: A Comparative Review* (B.C. Meggers, E.S. Ayensu and W.D. Duckworth, eds.). Smithsonian Press, Washington, D.C., pp. 49 – 61.
Smith, A.G. and Briden, J.C., 1977. *Mesozoic and Cenozoic Paleocontinental Maps.* Cambridge University Press, London.
Smith, D.L., 1985. Caribbean plate relative motions. In: *The Great American Biotic Interchange* (F.G. Stehli and S.D. Webb, eds.). Plenum Press, New York, N.Y., pp. 17 – 48.
Solem, A., 1979. Biogeographic significance of land snails, Paleozoic to Recent. In: *Historical Biogeography, Plate Tectonics, and the Changing Environment* (J. Gray and A.J. Boucot, eds.). Oregon State University Press, Corvallis, Oreg., pp. 277 – 287.
Specht, R.L., 1981. Biogeography of halophytic angiosperms (salt-marsh, mangrove and sea-grass). In: *Ecological Biogeography of Australia* (A. Keast, ed.). W. Junk, The Hague, pp. 577 – 589.
Spencer, J.W., 1895. Reconstruction of the Antillean Continent. *Geol. Soc. Am. Bull.,* 6: 103 – 140.
Springer, V.G., 1982. Pacific plate biogeography, with special reference to shore fishes. *Smithson. Contrib. Zool.,* 367: 1 – 182.
Stehli, F.G. and Wells, J.W., 1971. Diversity and age patterns in hermatypic corals. *Syst. Zool.,* 20: 115 – 126.

Stevens, C.H., 1977. Was development of brackish oceans a factor in Permian extinctions? *Geol. Soc. Am. Bull.,* 88: 133 – 138.

Stott, P., 1981. *Historical Plant Geography.* George Allen and Unwin, London.

Stromberg, P.C. and Crites, J.L., 1974. Specialization, body volume, and geographical distribution of Camallanidae (Nematoda). *Syst. Zool.,* 23(2): 189 – 201.

Sues, H.D. and Taquet, P., 1979. A pachycephalosaurid dinosaur from Madagascar and a Laurasia-Gondwanaland connection in the Cretaceous. *Nature,* 279: 633 – 635.

Sullivan, W., 1974. *Continents in Motion. The New Earth Debate.* McGraw-Hill, New York, N.Y.

Sykes, L.R., McCann, W.R. and Kafka, A.L., 1982. Motion of Caribbean plate during last 7 million years and implications for earlier Cenozoic movements. *J. Geophys. Res.,* 87: 10656 – 10676.

Szalay, F.S., 1975. Phylogeny of primate higher taxa. In: *Evolutionary History of the Primates* (F.S. Szalay and E. Delson, eds.). Academic Press, New York, N.Y., pp. 91 – 125.

Szalay, F.S. and Delson, E., 1979. *Evolutionary History of the Primates.* Academic Press, New York, N.Y.

Takhtajan, A., 1969. *Flowering Plants: Origin and Dispersal.* Oliver, Edinburgh.

Tarling, D.H., 1980. The geologic evolution of South America with special reference to the last 200 million years. In: *Evolutionary Biology of the New World Monkeys and Continental Drift* (R.L. Ciochon and A.B. Chiarelli, eds.). Plenum Press, New York, N.Y., pp. 1 – 41.

Theel, H., 1885. Report on the Holothuroidea. In: *Voyage of H.M.S. Challenger, Part II. Zoology,* 14: 1 – 290.

Thiede, J., 1980. Palaeo-oceanography, margin stratigraphy and paleophysiography of the Tertiary North Atlantic and Norwegian-Greenland seas. *Philos. Trans. R. Soc. London, Ser. A,* 294: 177 – 185.

Thiele, H., 1977. *Carabid Beetles in Their Environments.* Springer-Verlag, Berlin.

Thomson, K.S., 1969. The environment and distribution of Paleozoic sarcopterygian fishes. *Am. J. Sci.,* 267: 457 – 464.

Thorne, R.F., 1972. Major disjunctions in the geographic ranges of seed plants. *Q. Rev. Biol.,* 47: 365 – 411.

Thorne, R.F., 1973. Floristic relationships between tropical Africa and tropical America. In: *Tropical Forest Ecosystems in Africa and South America: A Comparative Review* (B.L. Meggers, E.S. Ayensu and W.D. Duckworth, eds.). Smithsonian Press, Washington, D.C., pp. 27 – 47.

Throckmorton, L.H., 1975. The phylogeny, ecology and geography of *Drosophila.* In: *Handbook of Genetics* (R.C. King, ed.). Plenum Press, New York, N.Y., pp. 421 – 469.

Thulborn, R.A., 1983. First mammal-like reptile from Australia. *Nature,* 306: 209.

Timm, T., 1980. Distribution of aquatic oligochaetes. In: *Aquatic Oligochaete Biology* (R.O. Brinkhurst and D.G. Cook, eds.). Plenum Press, New York, N.Y., pp. 55 – 77.

Tindale, N.B., 1981. The origin of the Lepidoptera relative to Australia. In: *Ecological Biogeography of Australia* (A. Keast, ed.). W. Junk, The Hague, pp. 957 – 976.

Towns, D.R. and Peters, W.L., 1980. Phylogenetic relationships of the Leptophlebiidae of New Zealand (Ephemeroptera). In: *Advances in Ephemeroptera Biology* (J.L. Flannagan and K.E. Marshall, eds.). Plenum Press, New York, N.Y., pp. 57 – 69.

Traub, R., 1972. The zoogeography of fleas as supporting the theory of continental drift. *J. Med. Entomol.,* 9: 584 – 589.

Trueb, L. and Tyler, M.J., 1974. Systematics and evolution of the Greater Antillean hylid frogs. *Univ. Kansas, Mus. Nat. Hist., Occas. Pap.,* 24: 1 – 60.

Tschudy, R.H., 1984. Palynological evidence for change in continental floras at the Cretaceous Tertiary boundary. In: *Catastrophes and Earth History* (W.A. Berggren and J.A. Van Couvering, eds.). Princeton University Press, Princeton, N.J., pp. 315 – 337.

Tuxen, S.L., 1978. Protura (Insecta) and Brazil during 400 million years of continental drift. *Stud. Neotrop. Fauna Environ.,* 13: 23 – 50.

Tyler, M.J. 1979. Herpetofaunal relationships of South America with Australia. In: *The South America Herpetofauna: Its Origin, Evolution, and Dispersal* (W.E. Duellman, ed.). *Mus. Nat. Hist., Univ. Kansas, Monogr.,* 7: 73 – 106.

Tyler, M.J., Watson, G.F. and Martin, A.A., 1981. The Amphibia: diversity and distribution. In: *Ecological Biogeography of Australia* (A. Keast, ed.). W. Junk, The Hague, pp. 1277 – 1301.

Underwood, G., 1976. A systematic analysis of boid snakes. In: *Morphology and Biology of Reptiles. Linnean Society Symposium, 3* (A. de A. Bellairs and C.B. Cox, eds.). Academic Press, London, pp. 151 – 175.

Usinger, R.L. and Matsuda R., 1959. *Classification of the Aradidae (Hemiptera-Heteroptera).* British Museum of Natural History, London.

Valentine, J.W. and Moores E.M., 1972. Global tectonics and the fossil record. *J. Geol.,* 80: 167 – 184.

van Andel, T.H., 1979. An eclectic overview of plate tectonics, paleogeography, and paleoceanography. In: *Historical Biogeography, Plate Tectonics and the Changing Environment* (J. Gray and A.J. Boucot, eds.). Oregon State University Press, Corvallis, Oreg., pp. 9 – 25.

van Andel, T.H., 1985. *New Views on an Old Planet.* Cambridge University Press, Cambridge.

van Andel, T.H., Thiede, J., Sclater, J.G. and Hay, W.W., 1977. Depositional history of the South Atlantic Ocean during the last 125 million years. *J. Geol.,* 85: 651 – 698.

Van Couvering, J.A.H., 1977. Early records of freshwater fishes in Africa. *Copeia,* (1): 163 – 166.

van Steenis, C.G.G.J., 1972. *Nothofagus,* key genus to plant geography. In: *Taxonomy, Phytogeography and Evolution* (D.H. Valentine, ed.). Academic Press, New York, N.Y., pp. 275 – 288.

van Steenis, C.G.G.J., 1979. Plant geography of east Malaysia. *Bot. J. Linn. Soc.,* 79: 97 – 178.

Van Valen, L., 1978. The beginning of the age of mammals. *Evol. Theory,* 4: 45 – 80.

Vanzolini, P.E. and Heyer, W.R., 1985. The American herpetofauna and the interchange. In: *The Great American Interchange* (F.G. Stehli and S.D. Webb, eds.). Plenum Press, New York, N.Y., pp. 475 – 487.

Vinogradova, N.G., 1979. The geographical distribution of the abyssal and hadal (ultra-abyssal) fauna in relation to the vertical zonation of the ocean. *Sarsia,* 64: 41 – 50.

Wake, D.B. and Lynch J.F., 1976. The distribution, ecology, and evolutionary history of plethodontid salamanders in tropical America. *Bull. Nat. Hist. Mus., Los Angeles County,* 25: 1 – 65.

Wake, D.B., Maxson L.R. and Wurst, G.Z., 1978. Genetic differentiation, albumin evolution and their biogeographic implications in plethodontid salamanders of California and southern Europe. *Evolution,* 32: 529 – 539.

Walker, K.F., 1981. The distribution of freshwater mussels (Mollusca: Pelecypoda) in the Australian zoogeographic region. In: *Ecological Biogeography of Australia* (A. Keast, ed.). W. Junk, The Hague, pp. 1233 – 1249.

Wallace, A.R., 1876. *The Geographical Distribution of Animals.* Macmillan and Co., London, 2 volumes.

Watson, J.A.L., 1981. Odonata (dragonflies and damselflies). In: *Ecological Biogeography of Australia* (A. Keast, ed.). W. Junk, The Hague, pp. 1141 – 1167.

Watt, J.C., 1975. The terrestrial insects. In: *Biogeography and Ecology in New Zealand* (G. Kuschel, ed.). W. Junk, The Hague, pp. 507 – 535.

Webb, S.D., 1978. A history of savanna vertebrates in the New World, Part II. South America and the Great Interchange. *Ann. Rev. Ecol. Syst.,* 9: 393 – 426.

Webb, S.D., 1985a. Main pathways of mammalian diversification in North America. In: *The Great American Interchange* (F.G. Stehli and S.D. Webb, eds.). Plenum Press, New York, N.Y., pp. 201 – 247.

Webb, S.D., 1985b. Late Cenozoic mammal dispersals between the Americas. In: *The Great American Interchange* (F.G. Stehli and S.D. Webb, eds.). Plenum Press, New York, N.Y., pp. 357 – 386.

Welty, J.C., 1979. *The Life of Birds.* Saunders College Publishers, Philadelphia, Pa., 2nd ed.

Wenz, S., 1969. Note sur quelques poissons actinopterygiens du Crétace supérieur de Bolivie. *Bull. Soc. Geol. Fr.,* 11: 434 – 438.

West, R.M. and Dawson, M.R., 1978. Vertebrate paleontology and the Cenozoic history of the North Atlantic region. *Polarforschung,* 48: 103 – 119.

White, S., 1983. Eastern Asian – eastern North American floristic relations: the plant community level. *Ann. Mo. Bot. Gard.,* 70: 734 – 747.

Whitmore, T.C. (ed.), 1981a. *Wallace's Line and Plate Tectonics.* Clarendon Press, Oxford.

Whitmore, T.C., 1981b. Wallace's Line and some other plants. In: *Wallace's Line and Plate Tectonics* (T.C. Whitmore, ed.). Clarendon Press, Oxford, pp. 70 – 80.

Wiley, E.O., 1976. The phylogeny and biogeography of fossil and recent gars (Actinopterygii: Lepisosteidae). *Miscel. Publ., Univ. Kansas, Mus. Nat. Hist.,* 64: 1 – 111.

Williams, W.D., 1981. Aquatic insects: an overview. *Monogr. Biol.,* 41: 1213 – 1229.

Wilson, J.T., 1963. *Continental Drift.* Sci. Am., pp. 1 – 16.

Wilson, J.T., 1973. Mantle plumes and plate motions. *Tectonophysics,* 19: 149 – 164.

Wolfe, J.A., 1981. Vicariance biogeography of angiosperms in relation to paleobotanical data. In: *Vicariance Biogeography: A Critique* (G. Nelson and D.E. Rosen, eds.). Columbia University Press, New York, N.Y., pp. 413 – 435.

Wood, C.E., Jr., 1972. Morphology and phytogeography: the classical approach to the study of disjunctions. *Ann. Mo. Bot. Gard.,* 59: 107 – 124.

Wood, R., 1976. *Stupendemys geographicus,* the world's largest turtle. *Breviora,* 436: 1 – 31.

Woodburne, M.O. and Zinsmeister, W.J., 1984. The first land mammal from Antarctica and its biogeographic implications. *J. Paleontol.,* 58: 913 – 948.

Ying, T., 1983. The floristic relationships of the temperate forest regions of China and the United States. *Ann. Mo. Bot. Gard.,* 70: 597 – 604.

Zinsmeister, W.J., 1979. Biogeographic significance of the Late Mesozoic and Early Tertiary molluscan faunas of Seymour Island (Antarctic Peninsula) to the final breakup of Gondwanaland. In: *Historical Biogeography, Plate Tectonics and the Changing Environment* (J. Gray and A.J. Boucot, eds.). Oregon State University Press, Corvallis, Oreg., pp. 349 – 355.

Zinsmeister, W.J., 1982. Late Cretaceous – Early Tertiary molluscan biogeography of the southern circum-Pacific. *J. Paleontol.,* 56: 84 – 102.

Zwick, P., 1977. Australian Blephariceridae (Diptera). *Aust. J. Zool., Suppl. Ser.,* 46: 1 – 121.

Zwick, P., 1981a. Plecoptera. In: *Ecological Biogeography of Australia* (A. Keast, ed.). W. Junk, The Hague, pp. 1171 – 1182.

Zwick, P., 1981b. Blephariceridae. In: *Ecological Biogeography of Australia* (A. Keast, ed.). W. Junk, The Hague, pp. 1185 – 1193.

SUBJECT INDEX

198

202